Geology of the country around Ilfracombe and Barnstaple

Inland from Barnstaple and Ilfracombe the countryside ranges from the mudflats and equable climate of the River Taw estuary, through rolling grassland with narrow wooded valleys, to the bleak open heather and grass moors of high Exmoor. The Bristol Channel coast to the north is marked by an impressive range of great sea cliffs.

This landscape is a reflection of the nature of its underlying rocks, born as soft sediments in seas and rivers 300 to 400 million years ago, and subsequently folded by compressive forces and subjected to long-continued processes of erosion. Thus rich alluvial soils of the estuary are bordered by country directly underlain by Carboniferous and uppermost Devonian shales and sandstones, locally calcareous, which support mixed pasture and arable farming, and within which a thin sequence of cherts is picked out by sharp ridges such as Codden Hill. Tough slates and sandstones of Devonian age have been shaped into the high moorland of Exmoor National Park and the great 'hog's back' cliffs of the north Devon and west Somerset coast.

Such a history accounts for the dominance of agriculture and tourism in the economy of the district. At a local level it explains the nature of materials used to build farmhouses and dry-stone walls, the presence of a variety of metalliferous ores which brought a mining industry to such places as Combe Martin and North Molton, the glacial origin of the clays from which Barum ware is made, and the lack of groundwater resources and consequent dependence on surface supplies. Even the old Taunton to Barnstaple railway followed a course within the area described that was dictated by a long-vanished Ice-Age river.

There is every reason to suppose that such natural constraints as governed the past will continue to govern the future.

Plate 1 Capstone Point, Ilfracombe
Folded and cleaved shales, siltstones and sandstones of the Kentisbury Slates. (A 13031)

BRITISH GEOLOGICAL SURVEY

England and Wales

E. A. EDMONDS,
A. WHITTAKER and
B. J. WILLIAMS

CONTRIBUTORS

Geophysics
J. M. C. Tombs

Mining
R. C Scrivener

Palaeontology
D. E. Butler
D. E. White
N. J. Riley

Geology of the country around Ilfracombe and Barnstaple

Memoir for 1:50 000 geological sheets 277 and 293,
New Series

Natural Environment Research Council

LONDON: HER MAJESTY'S STATIONERY OFFICE   1985

ISBN 0 11 884364 8

Bibliographical reference

EDMONDS, E. A., WHITTAKER, A., and WILLIAMS, B. J. 1985. Geology of the country around Ilfracombe and Barnstaple. *Mem. Br. Geol. Surv.*, Sheets 277 and 293, 97pp.

Authors

E. A. EDMONDS, MSc
B. J. WILLIAMS, BSc
British Geological Survey, St Just, 30 Pennsylvania Road, Exeter EX4 6BX

A. WHITTAKER, BSc, PhD
British Geological Survey, Keyworth, Nottingham NG12 5GG

Contributors

R. C. Scrivener, BSc, PhD
British Geological Survey, Exeter

D. E. Butler, BSc, PhD,
D. E. White, MSc, PhD
British Geological Survey, London

N. J. Riley, PhD
British Geological Survey, Keyworth

J. M. C. Tombs
Syntek, 98 St Martin's Lane, London WC2

Other publications of the Survey dealing with this and adjoining districts.

BOOKS

British Regional Geology
South-West England, 4th Edition, 1975

Memoirs
Bideford and Lundy Island (292)   1979
Bude and Bradworthy (307, 308)   1979
Chulmleigh (304)   1979

MAPS

1:625 000

Sheet 2   Geological
Sheet 2   Quaternary
Sheet 2   Aeromagnetic

1:250 000
Lundy (51N 06W)   Geology *in press*
Lundy (51N 06W)   Aeromagnetic anomaly
South-Western Approaches (Sheet 1)   Aeromagnetic

1:50 000
Sheet 292 (Bideford and Lundy Island) 1977
Sheet 293 (Barnstaple) 1982
Sheet 294 (Dulverton) provisional edition 1969
Sheet 307, 308 (Bude and Bradworthy) 1980
Sheet 309 (Chulmleigh) 1980
Sheet 310 (Tiverton) provisional edition 1969

1835   Geological Survey of Great Britain

150 Years of Service to the Nation

1985   British Geological Survey

Printed in the UK for HMSO
Dd 737379   C20   12/84

CONTENTS

PLATES

FIGURES

TABLES

PREFACE

The Ilfracombe and Barnstaple district is included in Old Series Geological Sheets 26 and 27, published in 1835 and revised in 1839. The ground was re-surveyed towards the end of the 19th century, again on the one-inch scale, by W. A. E. Ussher. Six-inch mapping of adjoining areas to the west, south-west and south, carried out between 1963 and 1973, has led to publication of the 1:50 000 sheets 292 (Bideford and Lundy Island), 307/308 (Bude and Bradworthy) and 309 (Chulmleigh). Sheet 294 (Dulverton), to the east of the present district, and Sheet 310 (Tiverton) to the south-east, have been issued as provisional editions, compiled from Old Series maps plus information from sources outside the Institute.

Six-inch survey of sheets 277 (Ilfracombe) and 293 (Barnstaple) was carried out between 1968 and 1977, mainly by Mr E. A. Edmonds, Mr B. J. Williams and Dr A. Whittaker, with small areas by Dr E. C. Freshney, Mr K. E. Beer and Mr J. E. Wright, under Mr G. Bisson as District Geologist. Messrs Edmonds, Williams and Whittaker are the authors of this memoir. The account of mining in the district has been written by Dr R. C. Scrivener, Mr J. M. C. Tombs has contributed a geophysical chapter, and Dr J. R. Hawkes has provided a petrological account of the Fremington dyke. Fossils have been collected by Dr D. E. Butler, Dr D. E. White, Mr S. P. Tunnicliff and by the surveyors, and identified by Dr Butler, Dr White, Dr M. A. Calver, Dr W. H. C. Ramsbottom and Dr N. J. Riley. Professor W. G. Chaloner has identified fossil plants, Dr C. H. C. Brunton has commented on some of the brachiopods, and Mr M. J. Reynolds has separated conodonts from the Ilfracombe Slates. Photographs were taken by Mr C. J. Jeffery and Mr H. J. Evans.

Thanks are due to the many landowners of the district for their helpful co-operation, to Kingston Minerals Ltd, Archibald Nott and Sons and E.C.C. Quarries Ltd for admission to working quarries, and in particular to Mr H. R. Thomas, agent for Fortescue Estates, for readily granting access to large areas of high Exmoor and for permission to camp there. We are also indebted to the Torquay Natural History Society, the Sedgwick Museum, Cambridge, and the trustees of the North Devon Athenaeum for the loan of fossils, and to Dr Gwyn Thomas of Imperial College for making available the Holwill Collection of fossils. Dr F. J. Rottenbury and Mr H. St L. Cookes have given freely of their local knowledge.

The maps to which this memoir relates, sheets 277 (Ilfracombe) and 293 (Barnstaple), are published at the scale of 1:50 000. The memoir has been compiled by Mr Edmonds and Mr Williams and edited by Mr Bisson.

G. M. BROWN
Director

British Geological Survey
Keyworth
Nottingham   NG12 5GG
11 June 1984

GEOLOGICAL SEQUENCE

The following rock formations are present in the district:

SUPERFICIAL DEPOSITS (DRIFT)

Recent and Pleistocene

Scree	Stony rubble on steep slopes
Peat	Thin peat of high Exmoor
Alluvium	Silts, clays and gravels of river courses and estuary
River terrace deposits	Silts and gravels
Pebbly clay and sand	Partly sorted sands, silts and clays with large and small pebbles
Glacial sand and gravel	Fine-grained gravels with sands
Head	Widespread mantle of stony silty sandy clay
Boulder clay	Clay with generally small stones

SOLID DEPOSITS

Carboniferous

Upper Carboniferous (Westphalian and Namurian)		*Generalised thickness*
		m
Bude Formation	Shales, siltstones and mudstones, with fairly massive sandstones	
Bideford Formation	Shales, siltstones and mudstones, with medium-grained feldspathic sandstones, in part laterally equivalent to the Bude Formation	300
Crackington Formation	Shales with thin turbidite sandstones	500
Lower Carboniferous (Dinantian)		
Codden Hill Chert	Shales with cherts and some lenticular limestones	250

Carboniferous

Transition Group: Pilton Shales	Shales with sandstones and calcareous sandstones and thin limy lenses	?500

Devonian

Upper Devonian		
Baggy Sandstones	Sandstones, siltstones and shales	450
Upcott Slates	Buff, grey-green and purple slates	250
Pickwell Down Sandstones	Purple and brown sandstones with shales	?1200
Morte Slates	Grey slates	?1500
Ilfracombe Slates		
Kentisbury Slates	Slates with sandstones	300
Mainly Middle Devonian		
Combe Martin Slates	Slates with limestones	120
Middle Devonian		
Lester Slates-and-Sandstones	Slates and sandstones	75
Wild Pear Slates	Slates with limestones	50
Hangman Grits	Sandstones with slates	1650
Middle and Lower Devonian		
Lynton Slates	Slates, siltstones and sandstones	350

Contemporaneous igneous rocks
Tuff

Intrusive igneous rocks
Lamprophyre (kersantite)

SIX-INCH MAPS

The following is a list of six-inch National Grid maps included, wholly or in part, in sheets 277 and 293, with the dates of survey. The surveyors are: Mr K. E. Beer, Mr E. A. Edmonds, Dr E. C. Freshney, Dr A. Whittaker, Mr B. J. Williams and Mr J. E. Wright.

SS 42 NE	Bideford Edmonds and Williams	1969
SS 42 SE	Weare Giffard Freshney	1968
SS 43 NE	Braunton Edmonds	1969
SS 43 SE	Instow Edmonds	1969
SS 44 NE	Mortehoe Edmonds	1969
SS 44 SE	Woolacombe Edmonds	1969
SS 52 NW	Newton Tracey Williams	1970–71
SS 52 NE	Codden Hill Williams	1970–71
SS 52 SW	Alverdiscott Freshney	1968
SS 52 SE	Atherington Freshney	1968
SS 53 NW	Marwood Edmonds	1969, 1974
SS 53 NE	Muddiford Edmonds	1975
SS 53 SW	Fremington Edmonds	1970
SS 53 SE	Barnstaple Edmonds	1970
SS 54 NW	Ilfracombe Whittaker	1977
SS 54 NE	Combe Martin Whittaker	1977
SS 54 SW	West Down Edmonds	1969, 1974
SS 54 SE	Berry Down Edmonds	1975
SS 62 NW	Chittlehampton Williams and Edmonds	1970–71
SS 62 NE	Filleigh Edmonds and Williams	1971, 1974
SS 62 SW	Umberleigh Beer	1968–69
SS 62 SE	Warkleigh Beer	1968
SS 63 NW	Bratton Fleming Edmonds	1976
SS 63 NE	Fullaford Williams	1976
SS 63 SW	Swimbridge Williams and Edmonds	1971, 1974
SS 63 SE	East Buckland Williams	1975
SS 64 NW	Trentishoe Whittaker and Edmonds	1975–76
SS 64 NE	Martinhoe Whittaker	1972, 1975–76
SS 64 SW	Arlington Edmonds	1976
SS 64 SE	Challacombe Edmonds and Whittaker	1975–76
SS 72 NW	South Molton Edmonds and Williams	1971, 1974
SS 72 NE	Newtown Williams and Edmonds	1971, 1975
SS 72 SW	Alswear Beer	1968
SS 72 SE	Bishop's Nympton Wright	1967
SS 73 NW	Muxworthy Williams	1976
SS 73 NE	Simonsbath Williams	1976
SS 73 SW	Heasley Mill Williams	1975
SS 73 SE	North Radworthy Williams	1975
SS 74 NW	Lynton Whittaker	1972–73, 1975
SS 74 NE	Brendon Whittaker	1972–73
SS 74 SW	The Chains Edmonds and Whittaker	1973, 1975–76
SS 74 SE	Brendon Two Gates Edmonds and Whittaker	1973, 1976
SS 75 SW	Blackhead Whittaker	1972
SS 75 SE	Foreland Point Whittaker	1972, 1975

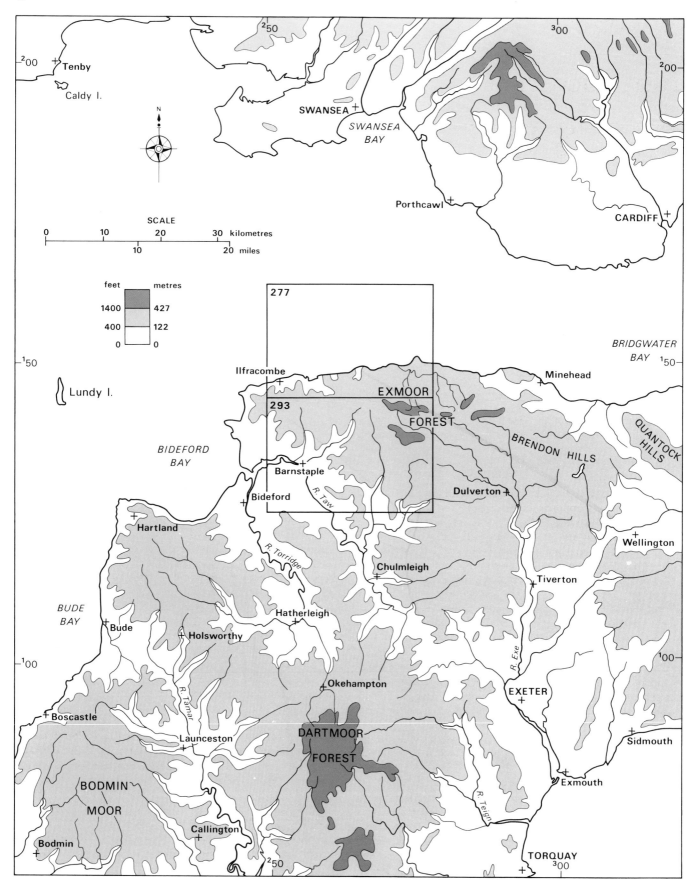

Figure 1 Sketch-map showing the position of the Ilfracombe and Barnstaple districts

CHAPTER 1

Introduction

POSITION, CLIMATE AND PHYSIOGRAPHY

The Barnstaple and Ilfracombe district (Figure 1) exhibits physical contrasts more striking than those of any other area of similar extent in southern England. It ranges from the mudflats, rich alluvial soils and equable climate of the River Taw estuary to the Chains of Exmoor, bleak grass moorland over 460 m (1500 ft) above sea level. Fremington, 5 km west of Barnstaple, lies at the western edge, Chittlehampton and South Molton near the southern and Simonsbath near the eastern edge. To the north is a range of magnificent sea cliffs; east of Little Hangman, for example, and within 500 m of the coast, the ground rises to over 335 m above sea level.

Low-lying tracts around Barnstaple and the estuary give way southwards to rolling countryside developed on Carboniferous strata, and northwards to the sharper relief of the Devonian rocks. In the north-east this more rugged landscape rises to high Exmoor, where broad summits fall gently to sudden deep-cut valleys. However, nowhere is it possible to get more than about 5 km from a road and there is not the same wildness, isolation and solitude as in the remoter parts of Dartmoor. The skylines remain gentle curves, undisturbed by rocky tors but occasionally broken by the silhouette of a group of the largest red deer in Britain. The high moors of The Chains, Hoaroak Hill and Exe Plain carry thin peat supporting deer grass and cotton grass. Whortleberry and scattered ling occur, together with some cross-leaved heath and a little bell heather, but in general, massed heather is peripheral to these central moors. Sedges and rushes characteristic of wet ground are common in the reclaimed pasture on the edges of the moor. Bracken is increasingly evident along the steep-sided river valleys. Thus the peatlands tend to remain visually dull and unchanging, in contrast to the autumn drifts of purple heather and the splashes of gold amid the dying bracken of the combes. The scenery of the district may therefore be summarised in a journey east-north-eastwards from Barnstaple. A varied and beautiful mixture of pasture and arable land near the estuary merges into grassland cut by narrow wooded valleys. This in turn gives way north-eastwards to exposed open heather and grass moors, grazed by red deer whose only cover is the stunted trees which survive in sheltered combes. Much of the Exmoor National Park is over 230 m above sea level, and some dip-and-scarp features and fault-line valleys reflect geological control of the topography.

Drainage in the west and south of the district is towards the River Taw. In the north-east it is eastward and south-eastward via the River Barle to the River Exe; and the Exe itself rises at Exe Head, north-west of Simonsbath, and flows slightly south of east towards Exford. An E-trending watershed north of and roughly parallel to these upper reaches of the River Exe lies generally within a kilometre of the river. To the north of it are headwaters of several short rivers and streams, including the East and West Lyn, which run off the northern flanks of Exmoor to the nearby Bristol Channel.

The local climate reflects relief and, in the low-lying western parts, the moderating influence of the sea. Barnstaple enjoys January and July temperatures of 7°C (44°F) and 16°C (61°F), and receives 1024 mm (40.3 in) of rain per year. Corresponding figures for Challacombe (260 m OD) are 3°C (37°F), 14°C (57°F) and 1675 mm (66 in), and for Chains Barrow (487 m OD) 1°C (34°F), 12°C (53°F) and 1950 mm (76.8 in). Precipitation at the last locality has ranged up to 2640 mm (104 in). The wettest month is October and the driest June. Snow does not lie for long around the estuary, but may persist on north-facing slopes of Exmoor until April. Sea mists are commonly blown up-river from the coast, and low cloud often envelops the high moors. The whole district, away from the lowest ground, is exposed to westerly winds from the Atlantic.

MAN AND INDUSTRY

Evidence of Stone Age man is scanty, and Five Barrows [732 368] is the only major Bronze Age site. Several Iron Age earthworks have survived, the largest being Mockham Down Camp [667 359] with a single wall, and Burridge Camp [569 352] and Shoulsbury Castle [705 391] with more than one wall. The Romans reached the district but left few traces, and although it has been suggested that some mining on Exmoor took place in Roman times the evidence is meagre. Westward-migrating Saxon peoples probably followed high-level trackways across Exmoor. They were here approaching the limits of their expansion, and seeking settlement and integration rather than conquest; a mixture of Saxon and British place names survives.

Agriculture and tourism dominate the district. The farming progression from the estuary to Exmoor is mixed arable and dairying, dairying, and finally sheep and beef, although the pattern varies greatly in detail according to soil and situation. There is little commercial forestry. Fishing, once an important if small industry of the northern ports, now plays an insignificant part in the local economy.

That good pasture can be maintained high on Exmoor, and even some arable farming practised in sheltered south-facing fields, is evident at 365 m OD around Simonsbath. However, until 1818 the high ground was used only for summer grazing for sheep from fifty or so parishes of lowland Devon, Cornwall and Somerset. From that year until 1898 the story of reclamation on Exmoor is the story of the Knight family, who bought 20 000 acres (8100 ha) around Simonsbath, built farmhouses, enclosed moorland with earth banks, beech hedges and shelter belts of conifers, laid out roads, constructed a canal eastwards from the impounded waters of Pinkworthy Pond, although for what purpose is not clear, and even participated in plans for a railway. Their farmsteads were typically built of local silty slate and sandstone, with Welsh roofing slate shipped into Lynmouth; they were sited on south-facing slopes and designed to enclose courtyards. Examples within the present district are Pinkworthy, Driver, Duredown, Cornham, Emmet's Grange and Wintershead.

The pattern of small-scale enclosure around Challacombe, on the edge of the moor, is medieval in character.

A variety of metalliferous ores has been mined in the past, including those of silver, copper, lead, zinc, iron and manganese, and the 'Bideford Black' culm seam extends into the district. Silver mining around Combe Martin, first recorded in medieval times, was carried on intermittently until finally abandoned in the 19th century. No mining now takes place within the district and none is likely in the foreseeable future. Most rock formations have been dug from time to time for local building stone, and a few quarries remain active in the harder sandstones, supplying material mainly for hardcore. Several quarries have been opened in limestones in the Combe Martin area, presumably mainly as sources of agricultural lime; all are now closed.

Barnstaple, with a population of 16 000, is a busy market town and the commercial centre of north Devon, at the lowest point at which the River Taw has been bridged. A bridge has existed here since the late 13th century; widening in and after 1834 has been confined to the upstream side, and it is possible to examine the underside of the old bridge from the pedestrian subway on the north bank. The town was an important port until the 19th century and retains many old stores and warehouses. The most important industry of the past was that of wool, but from about 1800 it gave way to cabinet making, and the manufacture of lace and gloves. Iron founding and agricultural engineering have flourished and declined.

Barum ware, made from the glacial clays of Fremington and once exported in large quantities to America, is still made at Brannam's pottery in Litchdon Street. This firm formerly owned the old North Walk Pottery near the castle, the site of which has yielded sherds and a potter's guild sign of the 17th century. Buff-yellow Marland bricks, of Tertiary clays from the Petrockstow Basin, are common in the buildings of the town, perhaps most prominent of all in the large cabinet and joinery works of Shapland and Petter at the southern end of the old Long Bridge across the Taw. Similar clays were used between 1859 and 1889 in John Seldon's factory in Alexandra Road, for the manufacture of long churchwarden pipes.

At least three shipyards have operated in Barnstaple. One, on the site of the Shapland and Petter joinery works, ended its production (of wooden ships) in 1887 with the ketch 'Emma Louise'. The adjoining concrete works replaced a shipyard which was active until the depression of the late 1920s. A third shipyard, and also some lime kilns, were situated where Taw Vale Parade now runs along the north bank of the river.

Queen Anne's Walk, perhaps the most interesting building in Barnstaple, is a reminder of the days of sea-borne trade It stands end-on to the river at the western end of the bus station and once had a quay alongside. Traders sealed a bargain by placing their money on the Tome Stone. The inlet was filled during construction of the railway to Ilfracombe. Several other wharfs survive, and serve small pleasure boats and the occasional barge carrying sand and gravel dredged from the estuary. Pilton Quay was reached along the River Yeo, which is so narrow and sinuous that vessels were commonly unable to turn around there. Wool, leather and bark were exported; coal, timber and iron came in. Ships plied Baltic, southern European, Irish and North American routes as well as British coastal runs. Rolle Quay, also alongside the River Yeo, carries old mills and warehouses reflecting busier days; some 100 000 tonnes of goods were handled there as recently as 1929.

Evidence of extensive water-borne trade reflects Barnstaple's position as a focus for local industry, both agricultural and manufacturing. The first railway (1848) in the district, for horse-drawn freight between Fremington and Barnstaple, was of standard gauge. Subsequent, initially broad-gauge, railways converged on the town from Exeter (opened 1854), Taunton (1873–1966) and Bideford (opened 1855). The Ilfracombe line opened as standard gauge in 1874 and closed in 1970. Only the Exeter line remains in use for all traffic; some freight, mainly ball clay, is brought from Meeth via Bideford. Most interesting of all, probably because of its narrow gauge of 1 ft 11 in (0.6 m), was the Lynton and Barnstaple railway (1898–1935). It started from alongside the Ilfracombe line at Barnstaple town station, traversed the site of the new civic centre, followed a still-traceable winding course along the valley of the River Yeo through Snapper and Chelfham, and ran thence via Bratton Fleming, Wistlandpound and Parracombe to Lynton. Chelfham Viaduct, over 40 ft (12.2 m) high with eight arches each 24 ft (7.3 m) across, stands intact except for the track, a monument to the consultant engineer Sir James Szlumper.

Most of the main roads in the district owe their origin to the Barnstaple Turnpike Trust, whose original turnpike act was passed in 1763 and renewed at intervals of 21 years. In 1879 the Trust set a granite quoin in the wall at the junction of the famous Butcher's Row and the High Street, and in the same and the following year erected 105 milestones alongside the local roads, each marked 'Barum' and showing the distance from this quoin. Now that the railways have ceased to serve the surrounding country areas, as traffic has swung back to the roads, the old turnpike network has been modernised and extended. New roads have been built, and one is planned to link Barnstaple and north Devon with the national motorway system. This link road will also serve South Molton (pop. 3000), a pleasant market town with a broad handsome High Street.

Ilfracombe (pop. 8000), Combe Martin (2000) and Lynton–Lynmouth (11 700) have grown from small fishing villages with local coastal trade, and small agricultural communities, into busy holiday and tourist centres.

Light industry is well established around Barnstaple, and urban and industrial development promises to expand alongside the Taw estuary. Little change may be expected in the countryside, where manufacturing industry will continue to be discouraged by poor local communications and distance from the great centres of population. Agriculture and tourism will continue to dominate the economy.

Major water supplies within the district all depend on surface storage. Small quantities of groundwater are available in the intergranular spaces and fractures of such formations as the Hangman Grits, Pickwell Down Sandstones and Baggy Sandstones, and in the fracture systems of more argillaceous rocks, but they are generally sufficient only for individual domestic or farm requirements. Water is brought to the area from Meldon Reservoir on northern Dartmoor, and 'local' water is impounded at Wistlandpound and Challacombe. The latter small reservoir was constructed in 1940 to supplement the supplies impounded at Slade, south-west of Ilfracombe, and the gravity feed to Ilfracombe was a considerable engineering achievement. Old pipelines are now being replaced by new, and pumps are taking over from gravity. A small reservoir on the River Yeo at Loxhore Bridge supplies water to Barnstaple. Geologically suitable sites exist for additional reservoirs, but the only major potential water resource within the district is the River Taw.

GEOLOGICAL HISTORY

The oldest rocks exposed in the district are the Lynton Slates, of Lower and possible Middle Devonian age. These sediments were laid down in shallow seas, probably with alternating periods of quiescence and active deposition. Local fault movements may have affected sedimentation. Possible beach deposits at the top of the Lynton Slates sequence herald the transition to the overlying Hangman Grits.

The Hangman Grits comprise massive, thick-bedded and thin-bedded sandstones with subordinate argillaceous material, mainly continental fluviatile deposits but showing an upward passage from deltaic and river sedimentation to a near-shore shallow-water marine environment supporting bivalves and gastropods. Probably land lay to the north or north-east and sea to the south or south-west, and the depositional area was subject to periodic sheet flooding. Within the marine transgression recorded in the higher Hangman Grits – Ilfracombe Slates transition there occurred minor transgressive and regressive events with local development of estuaries and fan deltas. Sediments of the Ilfracombe Slates mark the re-establishment of marine conditions. The Wild Pear Slates probably originated as locally calcareous muds and silts on delta slopes some distance south of the shore. The succeeding Lester Slates-and-Sandstones carry a little limestone. They give way to Combe Martin Slates which contain several coral and detrital limestones, reflecting the growth of coral reefs in the still waters of a shelf sea. The shallow-water Kentisbury Slates, mainly or wholly of Upper Devonian age, contain subordinate but distinctive sandstones marking more active deposition; they are succeeded by grey slates (Morte Slates) which indicate a return to the quieter waters of a shelf sea disturbed only sporadically by incursions of sand-bearing currents. An unstable shoreline, whose moves to north and south may be reflected in muddy sandy facies within the Ilfracombe Slates, migrated southwards at the end of Morte Slates deposition to give rise to a regime of shallow seas, lagoons and deltas in which accumulated the sands and muds now seen as Pickwell Down Sandstones, Upcott Slates and Baggy Sandstones. The base of the Pickwell Down Sandstones is marked in many places

by a band of tuff, locally several metres thick, whose volcanic source is unknown.

Shallow seas prevailed in late Devonian and early Carboniferous times, and the Pilton Shales, which span the junction of the two systems, yield a brachiopod-bivalve fauna in their lower part and a trilobite-goniatite fauna above. The overlying Lower Carboniferous strata comprise still-water shales, shallow-water cherts (Edmonds and others, 1968) and lagoonal limestone lenses. Deepening water then occupied a basin subject to periodic disturbance by sediment-carrying turbidity currents, and the muds and sands which accumulated therein now form the Crackington Formation. The topmost Carboniferous rocks are shallow-water deposits. The Bideford and Bude formations pass laterally one into the other; the former is of deltaic origin, the latter shallow-water marine, and the most obvious field distinction is between buff fine-grained and medium-grained feldspathic graded sandstones and grey-green fine-grained massive structureless sandstones.

Sedimentation was terminated by strong south–north compressive forces during the Variscan Orogeny, a major period of earth movement whose main phase took place in late Carboniferous times. The rocks of the district were intensely folded and reverse faulted, and given a characteristic east-south-easterly structural grain. The sandy formations resisted deformation and are commonly disposed in open folds. More argillaceous strata were squeezed into close folds which in the north of the district are locally overturned to the north. It is likely that the major NW-trending faults were initiated at this time. After the main compressive forces of the orogeny were spent, and during or after uplift of the area, there was probably a tensional phase of activity during which normal faulting took place. Some strike faults, perhaps including the Brushford Fault which separates Pilton Shales from Crackington Formation, may have moved during both compression and tension.

No New Red rocks are present in the district, although a scatter of pebbles may point to the presence at some time of conglomerates of that age. A cover of such rocks might also have constituted a source for the hematite which has locally been deposited in fractures. By Permian time a dramatic change of climate had taken place to arid or semi-arid conditions, and comparison with adjoining areas suggests that during the Permian and Triassic periods there was much intense erosion. One area of deposition was the present site of the Bristol Channel, and Triassic rocks are known at submarine crop very close to the north Devon coastline.

EAE, BJW, AW

The only intrusive igneous rock of the Barnstaple –Ilfracombe district is a small heavily weathered lamprophyre dyke exposed in Pilton Shales on the shore of the Taw estuary south of Penhill Point. It is probably related to the Permian lamprophyres found north and east of Dartmoor. No evidence remains of any younger solid rocks. By analogy with other parts of south-west England it is likely that the NW–SE wrench faults and complementary NE–SW fractures were active mainly in Tertiary times during the Alpine movements.

Early Tertiary drainage was to a south-easterly flowing River Taw, which underwent reversal. High-level river terraces point to an early Pleistocene river running westwards across the district almost along the line of the old Taunton to Barnstaple railway, and the disruption of this flow may date from the Middle Pleistocene Anglian glaciation. Ice of the next (Wolstonian) glaciation probably stood against the north Devon coast, and certainly advanced eastwards up the Taw estuary to Barnstaple, leaving boulder clay in its wake. It seems likely that local ice formed on Exmoor, disrupting the drainage locally and giving rise to incipient corries on some north-facing slopes. The fourth, third and second river terraces may date from Wolstonian times, and the first terrace from the succeeding (Ipswichian) interglacial period. Much head was formed by periglacial weathering during the final (Devensian) glaciation, when an ice front stood in South Wales and the northern part of the Bristol Channel. The great Exmoor flood of 1952 gave rise to a number of small landslips in the upper Exe valley.

HISTORY OF RESEARCH

De la Beche's classic *Report on the geology of Cornwall, Devon and west Somerset* (1839), the foundation of published research in south-west England, was followed in north Devon by structural studies by Etheridge (1867) and palaeontological notes from Hall (1867; 1876). However, the second major contribution to the geology of the district was that of Ussher, mapping for the Geological Survey in the last quarter of the 19th century. Ussher published a series of papers (1879; 1881; 1887; 1892; 1900; 1901; 1906), several of them with small-scale maps, and it was his work more than any other that facilitated the production of a geological map of north Devon, at ¾ inch to one mile (1:84 480), which accompanied an excursion report by Hamling and Rogers (1910). During the long period spanned by Ussher's work, Hicks (1891; 1896) wrote notes on the stratigraphy of north Devon, Hinde and Fox (1895) studied the Lower Carboniferous cherts, and Whidborne (1896; 1896–1907) a range of Upper Devonian fossils. Carboniferous rocks and their fossils were described by E. A. N. Arber (1904; 1907) and Rogers (1907; 1909; 1910; 1926).

In later years Paul (1937) and Goldring (1955) worked on the Pilton Shales, and Goldring (1971) made an exhaustive study of the Baggy Sandstones. Prentice (1960a) extended his studies of Carboniferous rocks inland to the Barnstaple district. Holwill (1961; 1963; 1964) and Holwill and others (1969) showed that Evans (1922) had failed to recognise repetition in the Ilfracombe Slates and that these strata could be divided into upper and lower divisions, in which the slates were associated with sandstones, separated by a middle division in which the slates contained limestones. House and Selwood (1966) summarised Devonian and Carboniferous palaeontology and stratigraphy in south-west England, Freshney and Taylor (1971; 1972) worked out an Upper Carboniferous stratigraphy in the cliffs south of Hartland Point, and the classification of the Carboniferous of south-west England was reviewed by Edmonds (1974).

Early recognition of glacial deposits at Fremington (Maw, 1864) was supplemented by Dewey's (1910) and Taylor's (1956) descriptions of associated erratics. Mitchell (1960; 1972), Stephens (1966), Bowen (1969), Edmonds (1972a) and Kidson and Wood (1974) have contributed to knowledge of the glacial history of the district, and Edmonds (1972b) has identified an early Pleistocene river course between South Molton and Barnstaple.

The geology of the district was related to a regional framework by Edmonds and others (1969; 1975).   EAE, BJW

CHAPTER 2

Devonian

GENERAL ACCOUNT

The oldest rocks of the district, the Lynton Slates, comprise slates, siltstones and sandstones of probable late Emsian age; they originated as shallow-water marine sediments. They are overlain by the Hangman Grits, probably mostly deposited in Eifelian times, which consist of sandstones and grits, with some coarser arenaceous rocks, interbedded with slates and silty slates, and which reflect continental, deltaic and fluviatile sedimentation. Probably the shoreline lay immediately to the south, the Old Red Sandstone land mass to the north over what is now South Wales, and the sediments accumulated on a low-lying coastal plain. Holwill and others (1969) envisaged flash floods, separated by quiescent intervals, and deposition as the overthickened top-set beds of a delta whose growth pushed the shoreline southwards. They noted the massive bedding of the sandstones, and the presence of channel and scour structures, conglomerates and cross-bedding, all pointing to the action of strong currents. The thicker bands of slate reflect more tranquil episodes. Towards the end of Hangman Grits deposition the shoreline migrated northwards and some shallow-water marine sedimentation occurred. Bivalves, brachiopods, gastropods and corals have been recorded.

The Givetian saw the origin of much of the slates, coral and detrital limestones, and sandstones that comprise the Ilfracombe Slates. The lowest of these beds, the Wild Pear Slates, comprise mainly slates in the south, but farther north are known to include siltstones, sandstones and some limestones, and may have been laid down in the deeper waters off the delta front. The succeeding Lester Slates-and-Sandstones contain much more sandstone and accumulated in shallow near-shore waters; in the north of the district they contain a few limestones made up of fossil debris.

The Combe Martin Slates in the northern part of the district contain coral and detrital limestones. Farther south they crop out as almost entirely slates, with subordinate silty and sandy bands. They mark a northward marine transgression and the establishment of a shelf sea over the whole of north Devon. Calm waters prevailed and coral banks grew. With renewed southward outgrowth of delta fronts came shallower waters and the deposition of muds and sands now seen as the slates and sandstones of the Kentisbury Slates. Holwill (1963) considered that the Givetian–Frasnian (Middle–Upper Devonian) junction occurred within the Combe Martin Slates but Orchard (1979) discussed the possibility that it lay within the Kentisbury Slates. A further transgression re-established a shelf sea and in this the Morte Slates originated as muds during the late Frasnian and early Famennian.

The rest of the Upper Devonian history of the district is the story of a slowly fluctuating shoreline with its associated on-shore, near-shore and off-shore sedimentation. The Pickwell Down Sandstones, current-bedded ripple-marked and with some interbedded shales, formed in rivers, coastal lakes, deltas and shallow sea waters. They pass upwards into variegated argillaceous rocks of the Upcott Slates which Goldring (1971) associated with swampy alluvial environments and shallow fresh-water lakes. The overlying Baggy Sandstones, predominantly shales and siltstones despite their name, mark a return to offshore shallow-water sedimentation. Shallow-sea deposition continued from Devonian into Carboniferous times, producing the Pilton Shales, of which a greater thickness is Devonian than is Carboniferous and which are therefore described in this chapter.

Lynton Slates

The Lynton Slates have variously been known as Linton calcareous slates, Linton grey beds, Linton Group and Lynton Beds. Sedgwick and Murchison (1836) divided the grauwacke of north Devon into lower and upper divisions and considered the upper division to be equivalent to Coal Measures strata. They subdivided the lower division into five groups which they later (1837) considered were younger than the lowest Cambrian, but older than the Silurian. Amongst the minor subdivisions of the grauwacke were the Linton grey beds, described and illustrated by De la Beche (1839). Subsequently the strata were known as the Linton Group (Phillips, 1841), but most commonly as the Lynton Beds, the term which appears in the Devonian part of the *Lexique Stratigraphique International*. The account that follows incorporates information drawn from unpublished notes kindly made available by Professor S. Simpson. In this account the strata are designated Lynton Slates. They are not referred to a lithostratigraphic hierarchy either as a Group or as a Formation, and it has not proved possible to subdivide them.

The Lynton Slates crop out at the foot of the cliffs from west of Heddon's Mouth [6464 4938] to Woody Bay [6775 4892]. Eastwards from there they form the cliffs to a place east of Lynmouth [7348 4954], and are traceable inland as a narrow ESE-trending strip from the Lynton area to the eastern boundary of the present district. Inland exposure is poor, and the cliff sections are easy of access in only a few places.

Nowhere is the base of the Lynton Slates visible. The exposed strata are thought to be between 300 and 400 m thick. Precise thicknesses are unknown because it is impossible to measure continuous vertical sections in sheer cliff faces and difficult to trace packets of strata laterally. Compounding the problems of access to certain critical areas are structural complications, mainly brought about by faulting, which make it difficult to trace individual sequences across major fractures.

The Lynton Slates comprise grey and dark grey silty slates with mudstones, siltstones and grey sandstones, usually fine- to medium-grained. In the vicinity of faults the strata are red-stained in places. In general, the lower part of the exposed sequence is sandy or silty, but with significant argillites, and the upper part is predominantly argillaceous except at the top, where sandstones, possibly beach deposits, herald the transition into the overlying Hangman Grits. It is difficult to locate a precise junction in the few places where the transition is accessible, but in general it is placed at the level where sandstones become the dominant lithology. The change from predominantly slaty to predominantly sandy lithologies produces a strong feature which is traceable inland; this feature is less clear where the Lynton Slates–Hangman Grits junction is a fault, in the extreme eastern part of the district.

The sandier parts of the sequence contain some very thin argillaceous wisps, commonly disrupted by burrows and cleavage. In silty lithologies the cleavage does not impart much fissility to the rock. Thin beds of dark grey limestone rich in crinoid debris are moderately common in the middle and higher parts of the Lynton Slates; also present are thin shell bands, interpreted by Simpson (1964) as lumachelles and in places represented by thin clastic limestones. One of the most characteristic features of the Lynton Slates is the occurrence of the trace-fossil *Chondrites* (Simpson, 1957). Foreshore exposures, especially in the Woody Bay area, reveal extensive bedding planes in argillaceous rock covered by branching tubes of sandstone referable to the ichnogenus *Chondrites*. Five specimens were illustrated by Simpson (1957).

The lowest, predominantly arenaceous, part of the exposed Lynton Slates is perhaps 200 m thick or more. Although the sequence comprises beds of sandstone and slate, commonly the argillaceous and sandy material is finely interlaminated, with individual laminae varying in thickness from less than a millimetre to one or two centimetres. The proportion of sandy argillaceous material is very variable; there are sequences up to 12 m or so thick which are essentially sandstone, but with numerous thin muddy partings or laminae. Bedding is in many cases indistinct, giving a massive appearance to the sandstones, and this is usually caused by *Chondrites* burrows which pervade the thick sandstone units. In the less sandy parts of the sequence cleavage is more obvious. Some sequences up to 2 m thick are predominantly argillaceous, but with thin sandstone partings or bands. Bed junctions are commonly diffuse. The uppermost 30 m or so of this lower arenaceous division differ from the underlying material in having discrete bands (from 0.15 to 1.5 m thick) of quartzitic sandstone, with no muddy partings alternating with argillaceous siltstones or silty shales. In addition there are very calcareous bands ranging from 1 cm to 0.3 m in thickness and a sandy shell bed (up to 3.6 m thick) with abundant modioliform bivalves. The quartzitic sandstones commonly contain the trace-fossil *Arenicolites*. The higher part of the Lynton Slates is perhaps 120 to 150 m thick, and comprises sequences of slates (up to 30 m) with intercalated sandstones similar in character to those of the lower division.

The precise stratigraphical position of the Lynton Slates has been controversial since the earliest days of geological work in north Devon. Their relationship to the Foreland Grits is not clear in the Ilfracombe district, where the junction is everywhere a fault. Williams (1837) considered that the Foreland Grits (see below) underlay the Lynton Slates and were thus the oldest strata exposed in the area. Sedgwick and Murchison (1840) equated the Foreland Grits with the Hangman Grits, which overlie the Lynton Slates. De la Beche (1839) clearly considered that the Lynton Slates lay stratigraphically between the Foreland Grits and the Hangman Grits, and Ussher and Champernowne (1879) favoured the same interpretation. Simpson (1951) considered that the Lynton Slates were the oldest exposed rocks of north Devon, a view confirmed by the present survey and some reconnaissance traverses in the area to the east.

The sediments of the Lynton Slates probably represent alternating periods of quieter water and more active deposition. Disturbed water is indicated by the concealed bed-junction preservation of *Chondrites* (Simpson, 1957), which indicates repeated phases of penecontemporaneous erosion, by the concentration of fossils in lumachelles and by the wave-rippled tops of some sandstones. Quieter periods of deposition are indicated by the fine laminae of argillaceous material in siltstone and sandstone lithologies. AW

Apart from the accounts of Phillips (1841) and Whidborne (1901), little descriptive palaeontology relating to the Lynton Slates has been published. In a more general work, Etheridge (1867) included a discussion of the fossils which were detailed in an accompanying table; the list was later added to by the observations of Whidborne (1901) and Hamling (1908). Where they expressed a view, these workers referred the strata to the Lower Devonian but Simpson (1964) thought that some at least might be early Eifelian in age.

The list of macrofossils given in Table 1 is based largely on material collected during the recent survey. Fossils from the Lynton Slates are usually distorted or otherwise poorly preserved and most horizons yield a very restricted fauna. In total though, the formation contains a variety of fossil forms among which bivalves, brachiopods, bryozoans and crinoid debris are predominant. Tentaculitoideans are locally abundant. Of the bivalves, forms of *Palaeoneilo* and of *Ptychopteria* (*Actinopteria*) are relatively common, and there are a few bands rich in *Modiomorpha*; *Carydium sociale* is found in large numbers at one locality, grouped in a manner reminiscent of the examples from the Ardennes described by Maillieux (1937). Among the brachiopods spiriferaceans are common; strophomenoids and orthoids are more thinly spread, though the remains of *Platyorthis longisulcata* are locally very numerous.

Considered as a whole, the macrofauna is suggestive of a level close to the Lower Devonian – Middle Devonian boundary, probably of late Emsian age; however, its constituent elements are scattered over a number of localities which cannot be related to a standard sequence and it is possible that strata of both Emsian and Eifelian age are present.

Only one of the samples taken from the Lynton Slates during the survey has yielded microfossils. Dr B. Owens reports carbonised plant spores including probable *Hystricosporites*, an Eifelian to Tournaisian genus, from this material, collected from the northern flanks of The Valley of Rocks [7086 4992].

Simpson (1964) concluded, largely on sedimentological grounds, that the Lynton Slates accumulated in a shallow sea, a view which is in general supported by their fossil content. Though a near-shore situation is not precluded by the fauna, the absence of lingulids, the relative importance of articulate brachiopods and, though Hamling (1908) reported some *Pteraspis* remains, the dearth of fish debris, together seem to militate against an environment open to direct fluviatile influence. The local occurrences of thick-shelled *Modiomorpha*, suited to a high-energy environment, may be indicative of off-shore shallows.   DEB

Table 1 Lynton Slates Fossils

Total fauna identified	Site	A	B	C	D	E	F	G	H	I	J	K	L	M	N
Plantae															
1 fragments indet.		.	1	.	1	1	.	.	.	.	.	1	.	.	.
Hydrozoa															
2 stromatoporoid?		.	.	.	.	.	.	.	.	.	.	.	.	.	.
Anthozoa															
3 solitary coral indet.		.	.	.	.	.	.	.	.	3	.	.	.	.	.
Bryozoa															
4 *Fenestella*		.	.	.	.	.	.	.	.	.	.	.	.	4	.
5 *Polypora*		.	5	.	.	.	.	.	.	.	.	.	.	5	.
6 fenestellid indet.		.	.	.	.	.	.	6	.	.	.	.	6	.	.
7 bryozoan indet.		.	.	.	7	.	.	.	.	.	.	.	.	.	.
Brachiopoda															
ORTHIDA															
8 *Platyorthis longisulcata*		.	cf.	.	.	.	.	.	.	.	.	.	8	8	.
STROPHOMENIDA															
9 *?Chonetes sarcinulatus*		.	.	9	.	.	.	.	.	.	.	.	.	.	.
10 *C. sordidus*		.	.	.	.	.	.	.	10	.	.	.	.	.	.
11 *Devonochonetes?*		.	.	.	.	.	.	.	.	.	.	.	.	.	.
12 *"Orthotetes hipponyx"* of Whidborne (1901)		.	.	.	.	.	.	.	.	12	.	.	.	.	.
13 chonetacean indet.		.	13	.	.	.	.	?	.	?	.	.	.	.	.
14 strophomenoid indet.		.	.	14	.	.	.	.	14	.	.	.	.	14	.
RHYNCHONELLIDA															
15 rhynchonellacean indet.		.	?	.	.	.	.	.	.	.	.	15	.	.	.
SPIRIFERIDA															
16 cf. *Alatiformia alatiformis*		16	.	.	.	.	.	.	.	.	.	.	.	.	.
17 *Brachyspirifer?*		.	.	.	.	.	.	.	.	17	.	.	.	.	.
18 cf. *Euryspirifer paradoxus*		.	18	.	.	.	.	.	.	.	.	18	.	.	.
19 *Fimbrispirifer?*		.	.	.	.	.	.	.	.	.	.	.	.	.	.
20 *Mauispirifer?*		.	.	.	.	.	.	20	.	.	.	.	.	.	.
21 *Paraspirifer*		.	.	.	.	.	.	.	.	.	.	.	.	.	.
22 *Retzia*		.	.	.	22	.	.	.	.	.	.	.	.	.	.
23 *Subcuspidella lateincisa*		.	.	.	.	.	.	23	cf.	cf.	.	.	.	.	.
24 athyridid?		.	.	.	.	.	.	.	.	.	.	.	.	.	.
25 spiriferacean indet.		.	.	25	25	.	25	.	.	.	.	.	.	.	.
Gastropoda															
26 *'Bellerophon'*		.	.	.	.	.	.	.	.	.	.	.	.	.	.
27 *Bucanella?*		.	27	.	.	.	.	.	.	.	.	.	.	.	.
28 *'Murchisonia'*		.	.	.	.	.	.	.	.	.	.	.	.	.	.
29 *Naticopsis?*		.	.	.	.	.	.	.	.	.	.	.	.	.	.
Cephalopoda															
30 *'Orthoceras'*		.	.	.	.	.	.	.	.	.	.	.	.	.	.
Bivalvia															
31 *Carydium sociale*		.	?	.	.	31	.	.	.	.	.	.	.	.	.
32 *Cimitaria* cf. *acutirostris*		.	.	.	.	.	.	.	.	.	.	32	.	.	.
33 *Cypricardella* cf. *rhomboidalis*		.	.	.	.	.	.	.	.	.	.	33	.	.	.
34 *Cypricardinia*		.	.	.	.	.	.	.	.	.	.	34	.	.	.
35 *Grammysia* cf. *ovata*		.	.	.	.	.	.	.	.	.	.	.	.	.	.
36 *Leptodesma?*		.	.	.	.	.	.	.	.	.	.	.	.	36	.
37 *'Leptodomus'* cf. *arcuata*		.	.	.	.	.	.	.	.	.	.	.	.	.	.
38 *'L.' lanceolata*		.	.	.	.	.	.	.	.	.	.	38	.	.	.
39 *Modiomorpha modiola*		.	.	.	.	.	.	.	.	39	cf.	39	.	.	.
40 *M.*		.	.	.	.	.	.	.	.	.	.	.	.	.	.
41 *'Myophoria'*		.	.	.	.	41	.	.	.	.	.	.	.	.	.
42 *Myilarca?*		.	.	.	.	42	.	.	.	.	.	.	.	.	.
43 *'Nuculana'*		.	.	.	.	43	.	.	.	.	.	?	.	.	.
44 *Nuculoidea ?trigona*		.	.	.	.	.	.	.	.	.	.	.	.	.	.

Table 1 continued	Total fauna identified	Site	A	B	C	D	E	F	G	H	I	J	K	L	M	N
45	*Palaeoneilo maureri*		.	.	.	.	cf.	cf.	.	.	.	.	45	.	.	.
46	*P.* cf. *primaeva*		.	.	.	.	.	46	46	.	46	.	46	.	.	.
47	*Paracyclas rugosa*		.	47	.	?	.	.	cf.	.	.	.	.	.	.	.
48	*?Prothyris plicata*		.	.	48	.	.	.	.	.	.	.	.	.	.	.
49	*P.?*		.	.	.	49	.	.	.	.	.	.	.	.	.	.
50	*Ptychopteria (Actinopteria) fasciculata*		.	.	.	.	.	.	.	.	.	.	.	.	.	.
51	*P. (A.) spinosa*		.	.	.	.	.	.	?	.	.	.	.	cf.	cf.	?
52	*P. (A.) sp. ?nov.*		.	.	.	52	.	.	.	.	52	.	52	.	.	.
53	*P. (A.)* cf. *musicata*		.	.	.	.	.	.	.	.	.	.	53	.	.	.
54	*?P. (A.)*		.	.	54	.	.	.	.	.	.	.	.	.	.	.
55	*Tancrediopsis* aff. *subcontracta*		.	.	.	.	.	55	.	.	.	.	.	.	.	.
	Trilobita															
56	*Phacops*		.	.	.	.	.	.	.	.	.	.	.	.	.	.
	Crinoidea															
57	columnals		.	57	57	57	.	57	57	.	57	.	57	.	57	57
	Tentaculitoidea															
58	*Dicricoconus?*		.	.	.	.	.	.	.	.	.	.	.	.	.	58
59	*Tentaculites*		.	59	.	?	?	.	.	.	.	.	.	.	.	?
	Trace fossils															
60	*Chondrites*		.	.	.	.	.	.	.	.	.	.	.	.	.	.
61	borings indet		.	.	61	.	.	.	61	.	.	.	.	.	.	.
			A	B	C	D	E	F	G	H	I	J	K	L	M	N

The forms listed have been identified from material collected during the survey, supplemented by material from museum collections. Occurrences at 14 selected sites (A to N) are shown, and relevant British Geological Survey specimen numbers and locality details are listed below. Some of the taxa listed have not been recorded from any of the sites selected for inclusion in the table.

Localities and specimen numbers

A   Heddon's Mouth, Trentishoe; cliff face [6533 4961]. DEA 1229–1248.
B   Exposures by and in floor of track to Woody Bay, Martinhoe [6764 4907 to 6767 4904]. DEA 1814–1873.
C   Lee Bay, Martinhoe; reef [6937 4919]. DEA 1200–1226.
D   Wringcliff Bay, Lynton; reef [7013 4975]. DEA 1144–1194.
E   Road cutting west-north-west of Barbrook [7060 4794]. DEA 1877–1930.
F   The Valley of Rocks, Lynton; south-west side [7029 4939]. DEA 1387–1428.
G   The Valley of Rocks, Lynton; south side [7065 4955]. DEA 1576–1634.

H   The Valley of Rocks, Lynton; north side [7097 4993]. DEA 1986 2022.
I   Hollerday Hill, Lynton; north-west flank [7118 4988]. DEA 1931–1975.
J   Reef north-west of Hollerday Hill, Lynton [7125 5006]. DEA 1651–1656.
K   Reef north of Hollerday Hill, Lynton [7148 5005]. DEA 1659–1804.
L   Road cutting, Myrtleberry Cleave, Lynton [7375 4871]. DEA 2109–2151.
M   Brendon; exposures in and by East Lyn River [7640 4802 to 7653 4812]. DEA 1051–1109.
N   Hillside south-east of Brendon [7698 4795]. DEA 1110–1126.

Hangman Grits

The Hangman Grits (Etheridge, 1867) are well exposed, but for the most part difficult of access, along the coast between Little Hangman and Woody Bay. The cliffs are up to 240 m high in places, with adjoining ground to the south rising to over 300 m. Inland the strata form a wide outcrop trending east-south-east in the west with regional dip to the south-south-west; in the east of the district the trend is about east–west and the southern margin bows southwards into north–south valleys in a manner consistent with a general southerly dip. The Foreland Grits are exposed eastwards from Lynmouth Bay.

The Hangman Grits succeed the Lynton Slates conformably. The sequence is mainly arenaceous and quartzose (Plate 2), but with alternating shales or slates and with minor intercalations of pebble beds or conglomerates. The sandstones range in colour from buffish grey to various shades of red, green and purple, and hematite staining is locally extensive. The proportion of shale or fine silt grade material is variable, depending upon stratigraphical level; laterally-persistent purple shales and siltstones are evident in the higher parts of the sequence.

The Hangman Grits have been variously known as Trentishoe Slates, Martinhoe group, Martinhoe beds and Hangman Sandstones. Strictly speaking, the term 'grit' is a misnomer when applied to these rocks. Gritty sandstones are present, but the bulk of the arenaceous deposits are sandstones of medium grain size. The beds are referred to as Hangman Grits in the Devonian part of the *Lexique Stratigraphique International*, and for historical reasons the Geological Survey has continued to use this term. Tunbridge

(1978) formalised the nomenclature and proposed a Hangman Sandstone Group, divided into several formations (Table 2) which are based upon earlier names and divisions of the strata.

Table 2   Hangman Grits stratigraphical sequences

Lane (1965)		Tunbridge (1978)	
	Thickness		*Thickness*
Hangman Grits	m	**Hangman Sandstone Group**	m
Stringocephalus Beds	100.58	Little Hangman Formation	c. 100
Sherrycombe Beds	82.30	Sherrycombe Formation	90
Rawn's Beds	100.58	Rawn's Formation	148
Trentishoe Grits	1066.80	Trentishoe Formation	c. 1250
		Hollowbrook Formation	70

The Foreland Grits, with type-area at Foreland Point [7541 5117], have also been referred to in the literature as the Foreland Sandstone and the Foreland Group. They comprise an intensely disturbed sequence of buffish grey, greenish grey, red, purple and green sandstones, alternating with cleaved siltstones or silty mudstones. Disruption is such that it is impossible to measure a continuous succession, but broadly speaking the beds are disposed in an anticline (the Lynton Anticline) whose axis follows roughly the ESE-trending hills which stretch from the Blackhead area of the coast through Countisbury Common and Kipscombe Hill to Old Barrow Hill. Plant remains, identified as *Calamites* and *Psilophyton*, have been recorded from the Foreland Grits (Evans, 1922; Evans and Stubblefield, 1929; Tunbridge, 1978), and Evans (1922) also noted the presence of 'obscure fish fragments'.

The question of the equivalence of the Hangman Grits and the Foreland Grits is fundamental to the understanding of the stratigraphy and structure of much of north Devon and west Somerset. Obvious lithological similarities have led to much discussion of possible correlation, but the evidence adduced has commonly been inconclusive. Ussher (1889) decided that the Hangman Grits and Foreland Grits should not be equated, although he was clearly uncertain; he remarked that if the correlation could be substantiated many difficulties experienced in mapping the area would cease to exist and the relations of the rocks would be much simplified. Of the four points that he thought were sufficient to negate the correlation, the first was the absence of conglomeratic beds in the Hangman Grits; these have since been recorded. His second point concerned the presence of a persistent

Plate 2   Sedimentary structures in the Hangman Grits at Heddon's Mouth
Convolute bedding in thin sandstones alternating with siltstone laminae. (A 13020)

horizon of quartzose grits with associated red slaty beds in the Hangman Grits but not in the Foreland Grits; in fact these Rawn's Formation beds (see below) are traceable with certainty only between the Rawn's Rocks area and Trentishoe Down; east-south-east of there, along the strike, are intermittent, usually poor, exposures, and scattered field fragments, of purple slate and pebbly or gritty beds commonly associated with distinctive green sandstones; similar green sandstones are present in the faulted southern limb of the Lynton Anticline, commonly associated with purple slates. Ussher's third point was the occurrence of *Natica* and *Myalina* in the Hangman Grits and their absence from the Foreland Grits; however, only the topmost 200 m or so of the Hangman Grits yield shelly fossils, and these sparingly, and it is likely that most of these beds have been faulted out along the major fracture which separates Foreland Grits from Lynton Slates. Ussher's fourth point was the contrast between the dip-and-scarp topography of the Hangman Grits and the 'dome-shaped or conical features' of the Foreland Grits; probably the latter topography reflects more intense folding and faulting, a possibility that Ussher did not overlook. Thus all of Ussher's (1889) objections can be accommodated. The quartzose sandstones of the Foreland Grits are lithologically similar to those of the Hangman Grits, and contain many parallel-laminated and convolute-bedded strata like those of the Trentishoe Formation (see below). Also, beds on the northern limb of the Lynton Anticline, exposed on the coast north-east and east of Upper Blackhead [7453 5032], closely resemble those of the Trentishoe Formation.

Evans (1922) established subdivisions of the Hangman Grits in the Combe Martin area which have survived with modification to the present time. His suggested succession was as follows:

	Thickness m
Hangman Grits	
Little Hangman Beds	121.92
Rawn's Beds (or Shales)	60.96
Trentishoe Grits	914.40

In 1929 (*in* Evans and Stubblefield, 1929) he introduced his 'beds of Sherry Combe', overlying the Rawn's Beds and containing casts of fossils, and the Stringocephalus Beds, lying 'presumably' above the Sherrycombe Beds. Equivalent strata in the Quantock Hills were investigated by Webby (1966a) and the Hangman Grits of the type area were described by Lane (1965) and Tunbridge (1976; 1978), whose stratigraphical sequences are given in Table 2. Tunbridge's divisions are recognisable over much of the area and are followed below.

Accurate thicknesses of the Hangman Grits are difficult to determine, and this is particularly true of the Trentishoe Formation, where problems of access are compounded by faulting and bedding-plane slip. Published thicknesses range from 1097 m (Evans, 1922) through 1350 m (Lane, 1965) to 1658 m (Tunbridge, 1978). Tunbridge's figure includes a formation (the Hollowbrook Formation) not distinguished by Evans or Lane, and is accepted here as the most reliable. However, calculations based on width of outcrop and inclination of strata in areas least disturbed by folding and faulting give a different order of thickness, ranging from 2000 m at an average dip of 30° to over 2500 m at 35°.

HOLLOWBROOK FORMATION

The Lynton Slates are overlain by the Hollowbrook Formation, consisting of between 70 and 90 m of beds in which sandstone is predominant, but with mudstone intercalations. Apart from the trace fossil *Arenicolites*, the formation is poorly fossiliferous, although tentaculitids are present (Tunbridge, 1978). Quartzitic grey mudstone is the dominant lithology, and beds are commonly parallel-laminated, with a few cross-bedded intercalations. Tunbridge (1978) recognised ladder ripples and reactivation surfaces, and uncommon channel forms filled with flaser and lenticular beds; he interpreted the transition from Lynton Slates to Hangman Grits in terms of an approaching wave-dominated shoreline, with well-developed beach facies.

TRENTISHOE FORMATION

The non-marine Trentishoe Formation comprises a thick monotonous sequence of buffish or pinkish grey quartzose sandstones with shale and silty shale intercalations. The sandstones are fine to medium grained, parallel-laminated, and commonly display convolute bedding. Scattered plant remains have been recorded. In the fine-grained sediments Tunbridge (1976) noted climbing ripples, ripple-laminations, plane beds and suncracks. He interpreted the sediments of the formation in terms of periodic sheet flooding.

RAWN'S FORMATION

The Rawn's Formation is about 150 m thick and contains the coarsest-grained arenaceous sediments, ranging from sandstones to gritty sandstones, pebble beds and fine conglomerates, together with purple slates. Evans (1922 *in* Evans and Stubblefield, 1929) recorded a *Coccosteus* scale and plant remains from his Rawn's Beds. According to Tunbridge (1976; 1978) the pebble assemblage contains angular exotic clasts (including igneous rocks), possibly derived from a nearby northerly source. The deposits are non-marine and are succeeded by transgressive muddy sediments.

SHERRYCOMBE FORMATION

The Sherrycombe Formation is about 90 m thick and includes lithologies ranging from grey lenticular mudstones to cross-bedded sandstones. Purplish grey sandstones contain numerous casts of fossils. Lane (1965) cited '*Myalina, Naticopsis, Bellerophon,* a spiriferid, Tentaculites, Thamnopora, Caulostrepsis, an *Arthrodire* plate, polyzoans and plant remains' [sic] from the formation. Tunbridge (1976; 1978) reported upwards-coarsening sequences with uncommon shelly horizons. He interpreted the sediments in terms of minor transgressive and regressive events against a general background of overall transgression, with the development of estuaries or possibly fan-deltas.

LITTLE HANGMAN FORMATION

The Little Hangman Formation, broadly equivalent to Evans's (*in* Evans and Stubblefield, 1929) and Lane's (1965)

Stringocephalus Beds, is probably of the order of 100 m thick. It comprises alternations of grey silty shales and siltstones with cross-bedded grey and pinkish grey fine-grained sandstones. The topmost beds, intensely folded and faulted at Little Hangman [5820 4790], show sandstone content decreasing upwards. The junction with the Ilfracombe Slates, though nowhere clearly seen is probably transitional and conformable. The Little Hangman Formation has yielded *Myalina* (see p. 10), *Naticopsis* and bryozoans (Lane, 1965). Tunbridge (1978) envisaged marine deposition with tidal influences.

Inland, exposure is poor, but mapping is facilitated by recognition of topographic features related to the solid geology. Such features are well displayed in the Trentishoe area, the Ilkerton Ridge – Furzehill Common areas and intermittently between, and at Withycombe Ridge. Four indistinct features associated with the combined Hollowbrook and Trentishoe formations trend east-south-east from the coast towards The Chains area, and thence eastward to the eastern edge of the district and beyond. These features are not readily correlatable with particular packets of strata; the most probable alternative explanation is that they are due to strike faulting, but it seems unlikely that such persistent fault belts are present within the Hangman Grits. Furthermore, the features accord with dips measured at outcrop in such a way as to suggest dip-and-scarp ridges related to the stratigraphy, with northerly-facing scarp slopes and southerly-inclined dip slopes.

The highest parts of the succession, including the lower, sandier, beds of the Little Hangman Formation, the Sherrycombe Formation and probably part of the Rawn's Formation, form a strong dip-and-scarp feature which is traceable with certainty from Little Hangman, via Great Hangman, Holdstone Down, Trentishoe Down, Heale Down and South Down to the vicinity of Parracombe. The feature can be followed farther east to Cannon Hill and the Ilkerton Ridge area and beyond, but there it is less obvious. Short steep north-facing or north-east-facing scarp slopes, and long gentle southerly or south-westerly dip slopes, accord with directions of dip measured at outcrop. However, the long dip slopes are much less steep than the measured dips, which range from 25° to 45° and perhaps average 30° to 35°. Slight changes of strike of these features are also confirmed by local dip measurements at outcrop. Thus southerly dips at Little Hangman give way to south-south-westerly dips from Girt Down to Holdstone Down and Trentishoe Down, which last two ridges are offset, one from the other, by faulting. From Heale Down to Cheriton Ridge the dip locally approaches south-westerly. Much of the Rawn's Formation appears to be associated with a topographic low which can be followed eastwards only as far as Martinhoe Cross [6814 4651], although traces of pebble beds are recognisable still farther east.   AW

Little good direct palaeontological evidence of the age of the Hangman Grits is available. However, in view of their position above the Lynton Slates and below the Ilfracombe Slates, they are largely if not entirely of Middle Devonian age, probably mainly Eifelian with some Givetian strata in their upper part. Evans (1922) first placed them in Middle, rather than Lower, Devonian, because they had yielded a scale of *Coccosteus*, because a limestone near the top (now

regarded as basal Ilfracombe Slates) had yielded what was thought to be a brachiopod closely related to *Stringocephalus burtini* (but is now identified as the bivalve *Myalina*), and because there was no stratigraphical break between the Hangman Grits and the Ilfracombe Slates.   EAE

Ilfracombe Slates

The Ilfracombe Slates crop out on the coast between Ilfracombe and Combe Martin as slates with subordinate sandstones and limestones, and they extend east-south-eastwards in a continuous but narrowing outcrop. The names Ilfracombe Group (first used by Phillips, 1841) and Ilfracombe Beds have also been applied to these rocks. Evans (1922) made the first attempt to elucidate the detailed stratigraphy, and his Wild Pear Beds and Lester Series survive as Wild Pear Slates and Lester Slates-and-Sandstones. His proposed succession, which did not take into account some repetition by folding and faulting, was as follows:

	Thickness m
Woolscott Cleave Slates: Coarse arenaceous slates	76.2
Slates with subordinate grits and limestones	304.8
David's Hole Beds: Striped calcareous slates with intercalated limestones (thickness uncertain on account of folding and faulting)	76.2
Red Limestone Series: Striped calcareous slates, bituminous limestone, massive limestone weathering red, and other slates and limestones (thickness uncertain on account of faulting)	108.2
Jenny Start Beds: Flaggy limestone	9.1
Slates and limestones of Oakestor Bay	41.2
Newberry Beds: Slates and limestones	158.5
Lester Series: Slates with grits and lenticular limestones; calcareous slates and limestones; grits and slates	320.0
Wild Pear Beds: Calcareous slates and lenticular limestones	121.9

Evans later (*in* Evans and Stubblefield, 1929) modified this 1216-m sequence to exclude the 'Slates and limestones of Oakestor Bay', and to change the name of the 'David's Hole Beds' to the 'David's Hole and Hagginton Beds', a clear indication that he recognised the equivalence of the strata exposed in Sandy Bay (David's Hole) [5696 4738] and in Hele Bay (West Hagginton Beach) [5419 4830]. The Red Limestone Series and the David's Hole Beds are now considered to be equivalent (Holwill, 1963).

Shearman (1962) recognised a three-fold division of the slates, with the upper and lower parts characterised by sandstones and the middle part containing most of the limestone. This approach was developed by Holwill (1963) and Holwill and others (1969), who discerned repetition of folding and faulting, identified marker limestones, and to whom most of the modern classification may be attributed. The present survey has shown that it is possible to follow stratigraphical divisions for considerable distances inland.

Classifications by Evans and Holwill are compared in Table 3 with that adopted by this memoir. Holwill and others (1969) Stringocephalus Beds are the *Myalina*-bearing beds, now referred on lithological grounds to the Little Hangman Formation of the Hangman Grits. The sediments comprising the Ilfracombe Slates record the completion of the change to marine deposition first indicated by the Little

Table 3 Ilfracombe Slates stratigraphical sequences

BGS (this account)		Holwill and others (1969)		Evans (1922) Evans and Stubblefield (1929)	
Kentisbury Slates	Slates with sandstones and thin limestones	Upper division	Mainly arenaceous beds	Slates with grits and limestones. Possibly includes some Woolscott Cleave Slates and David's Hole Beds = Red Limestone Series	
Combe Martin Slates	Slates with thin limestones and sandstones			David's Hole Beds = Red Limestone Series	
	David's Stone, Limestone	Middle division	David's Stone Limestone		
	Slates and sandstones		Slates with thin limestones		
	Combe Martin Beach Limestone		Combe Martin Beach Limestone		
	Slates, sandstones and thin limestones		Slates with thin limestones		
	Jenny Start Limestone		Jenny Start Limestone	Jenny Start Beds = Newberry Beds	
	Slates		Slates with thin limestones		
			Rillage Limestone		
Lester Slates-and-Sandstones	Rillage Limestone				
	Slates with sandstones and thin limestones	Lower division	Lester beds	Mainly arenaceous beds	Lester Series = slates and limestones of Oakestor
				Holey Limestone	
				Mainly arenaceous beds	
Wild Pear Slates			Wild Pear Beds	Wild Pear Beds	
			Stringocephalus Beds		

Hangman Formation (= Stringocephalus Beds), and there is no important sedimentological break at the Hangman Grits–Ilfracombe Slates junction.

Detailed stratigraphy is difficult to elucidate, except perhaps within the Lester Slates-and-Sandstones and the Combe Martin Slates, owing to the degree of deformation of the predominantly argillaceous rocks. Holwill's (1963) unravelling of the structure demonstrated that there are only two main limestones in a sequence about 106 m thick, compared with Evans's (1922) estimate of 393 m. Probably the Ilfracombe Slates have a maximum stratigraphical thickness of about 545 m compared with Evans's (1922) figure of 1216 m. AW, EAE

The fossils of the Ilfracombe Slates are in general either badly preserved or difficult to separate from the matrix. As a result, apart from corals, which can be determined from sections, they are poorly known. Very few featured in the descriptive works of the 19th century. Phillips (1841) recorded occurrences of a small number of species in the formation, though none exclusive to it, and isolated examples were figured by Milne Edwards and Haime (1850–1855) and by Davidson (1864–1865). Fossil lists have appeared from time to time, notably those of Etheridge (1867) and Valpy (1867); the latter also provided general locality details. In recent years corals have been described and figured by Holwill (1964a; 1968).

Macrofossils from the Ilfracombe Slates are listed in Table 4, where occurrences at 17 specific sites are shown. The list is based upon examination of material gathered during the survey and of fossils housed in a number of collections, largely in museums. The locality details of most museum specimens are not precise however.

Since the argillites are highly cleaved, almost all the fossils known are from the limestones, which account for only a small part of the formation. Even within the limestones, fossils are commonly only visible in section, owing to recrystallisation. For these reasons the list probably exaggerates the proportion of corals in the total fauna. Numerous samples of the limestones were taken during the survey and processed for conodonts; the few faunas obtained were described by Orchard (1979). DEB

Etheridge (1867) assigned the Ilfracombe Slates to the Middle Devonian but later authors, eg. Evans (1922) and Holwill (1963), while regarding the lower beds as of that age, considered the formation to range into the Upper Devonian. Holwill considered that differences in the coral faunas of the Jenny Start Limestone (compound and solitary rugose corals with subordinate tabulate corals), the Combe Martin Beach Limestone (small solitary rugose corals with subordinate tabulate corals) and the David's Stone Limestone (predominantly tabulate corals with subordinate small solitary rugose corals) had a chronological significance. He noted that a coral association similar to that of the Combe Martin Beach Limestone has been reported from the upper Middle Devonian of Bohemia (Prantl, 1938) and from Australia in strata of probable early Frasnian age (Hill, 1939). He considered that the coral faunas of the Combe Martin Beach and David's Stone limestones had features in common, and, on balance, were probably of early Frasnian age, but that both contrasted with that of the Jenny Start Limestone of late Givetian age. However, Webby (1966b) argued that the faunal differences of the limestones reflected different environments of accumulation. This interpretation is preferred here and is supported by the work of Fedorowski (1965), who concluded that the absence of colonial corals and stromatoporoids from an association of small solitary corals of Middle Devonian age in the Holy Cross Mountains, Poland, was an indication of a depth of water beyond the tolerance of such forms.

The coral fauna of the Jenny Start Limestone indicates a Givetian age, as Holwill (1963) observed. Of the typical Givetian forms, *Disphyllum aequiseptatum* and *Thamnophyllum caespitosum* are abundant. Tsien (1970) reported that in the Dinant Basin, Belgium, *D. aequiseptatum* is confined to the upper part of the Givetian. Scrutton (1968) reported *T. caespitosum* from the mid-Givetian limestones of south Devon.

No stratigraphic interpretation can be deduced from the occurrence of small lindstroemiids in the Combe Martin Beach Limestone. However, in the restricted tabulate coral fauna of the David's Stone Limestone, *Thamnopora cervicornis* is fairly well represented, a species which according to Lecompte (1939) is restricted to the Givetian in the Dinant Basin. Furthermore, the Holwell Limestone, which is the possible correlative of the David's Stone Limestone in the Quantock Hills to the east of the present district, contains a varied coral fauna of Givetian aspect (Webby, 1966b, p. 331).

Thus the coral faunas of the limestones of the Combe Martin Slates provide no evidence of a Frasnian age (see also Webby, 1966b), and on this basis the boundary between the Middle and Upper Devonian is considered to be at the top of the Combe Martin Slates or within the overlying Kentisbury Slates. Furthermore, Orchard (1979), having investigated conodont faunas, also concluded that the limestones of the Combe Martin Slates should be assigned to the Middle Devonian.   DEW

WILD PEAR SLATES

The Wild Pear Slates are exposed at Wild Pear Beach [5822 4778], faulted against strata of the underlying Little Hangman Formation. They comprise strongly overfolded and cleaved silvery grey slates with subordinate, thin sandstones, siltstones and limestones generally ferruginous and iron-stained. The trace-fossil *Chondrites* is present, but the shelly fauna is poor. Structural complications obscure the junction with the Lester Slates-and-Sandstones, but it is probably gradational and is taken below the lowest thick sandstone. The intense folding makes it difficult to estimate thickness, but the formation is much thinner than the 122 m suggested by Evans (1922), and possibly little more than 30 m. The sediments indicate marine deposition with a shoreline to the north.

Inland from Wild Pear Beach the slates strike east-south-east to the vicinity of Voley [640 460], and thence more south-easterly towards Parracombe. Discontinuous dip-and-scarp features between North Challacombe [5874 4798] and Holdstone Farm [6196 4698] probably reflect structurally

displaced harder strata in this generally slaty and incompetent succession. Exposure is poor, but field surface debris and scattered outcrops show that the lithologies visible at the coast persist inland. The outcrop narrows south-eastwards to Parracombe, and in the vicinity of Holworthy [6815 4409] the Wild Pear Slates are faulted out.

LESTER SLATES-AND-SANDSTONES

The Lester Slates-and-Sandstones are exposed in strike section at Lester Cliff [5792 4758] and Combe Martin Beach [5752 4761]. Further coastal exposures occur between Golden Cove [5656 4764] and Rillage Point [5420 4866]. The succession comprises slates, sandstones (locally cross-bedded), gritty sandstones, siltstones and mudstones (commonly with abundant *Chondrites*), and thin, crinoidal, shelly and rather nodular limestones. The Holey Limestone, near the top, is up to 0.8 m thick, dark grey and with mudstone partings; it yields corals and brachiopods and is named after its characteristic large solution hollows.

The 35- to 40-m sequence at Lester Cliff (p. 28) is repeated in many places on the coast in normal and inverted succession because of overfolding and thrusting. Where, as is commonly the case, only part of the formation is visible, correlation is nevertheless generally possible, and it is likely that the succession at Lester Cliff shows about the total thickness of the formation on the coast.

Inland from Lester Cliff, the Lester Slates-and-Sandstones form strong dip-and-scarp features, with steep north-facing scarp slopes and gentle south-south-west-facing dip slopes. Scattered outcrops show the same lithologies as those on the coast. The disposition of the features and the width of outcrop suggest that in the vicinity of Combe Martin the deep valleys between features are occupied by east–west strike faults. Mapping suggests that the deformation observed on the coast persists inland at least as far as Verwill [622 465]; farther south-east, the strike swings from east-south-east to south-east, the outcrop narrows, only one mappable feature is present in the formation and the structure may be less complex.

Lester Slates-and-Sandstones crop out in a valley [649 440] north-east of Hollacombe and at Rowley Cross [662 441]. Their southern margin bears a relationship to topography which suggests a formational dip of about 14° to slightly west of south and a thickness, assuming no repetition, of about 100 m. The beds terminate against a NW-trending fault which follows the valley south and south-east of Highley [6785 4391].

The only fossil that Evans (1922) named from the Lester Slates-and-Sandstones was *Tentaculites*. However, he noted brachiopods and gastropods in the Holey Limestone and, in the slates 'cylindrical forms, sometimes referred to as 'fucoids', which are more arenaceous than the surrounding slates, and appear to represent worm-burrows'; clearly, these are *Chondrites*, which occur in abundance. Holwill and others (1969) noted that the thin detrital limestones of these beds were formed from fossil debris, mainly brachiopods and bryozoans; he also recorded the coral *Thamnopora*. The Rillage Limestone, at the base of Holwill and others (1969) Middle Ilfracombe Beds, contains brachiopods, crinoids, orthocones, bryozoans and algae, together with uncommon fish fragments, ostracods and small gastropods, and is probably the lateral equivalent of the Holey Limestone, here placed with associated thin limestones at the top of the Lester Slates-and-Sandstones. Orchard (1979) referred conodonts from near the top of the Lester Slates-and-Sandstones to *Icriodus* aff. *difficilis*.

The sediments point to shallower seas and a higher energy environment than that of the Wild Pear Slates, alternating with periods of quieter, deeper water sedimentation. AW, EAE

COMBE MARTIN SLATES

The Combe Martin Slates, comprising slates with three distinctive limestone beds, and subordinate thin sandstones and siltstones, are exposed on the coast east of Ilfracombe and in Combe Martin Bay. They are equivalent to Holwill and others (1969) Middle Ilfracombe Beds, and include the more important limestone horizons of the Ilfracombe Slates. The succession elucidated by Holwill (1963) and Holwill and others (1969) is generally confirmed here. However, the base of the Combe Martin Slates is taken at about 10 m above the top of the Holey Limestone, which is thought to be the lateral equivalent of Holwill's (1964b) Rillage Limestone; the 10 m of strata above the limestone contain sandstones and siltstones with abundant *Chondrites* and are assigned to the Lester Slates-and-Sandstones. About 30 m of slates with thin limestones above the David's Stone Limestone are included in the Combe Martin Slates, a generalised sequence of which is as follows:

	Thickness
	m
Slates, with thin limestones and sandstones	30.00
David's Stone Limestone	9.00
Slates and sandstones	34.00
Combe Martin Beach Limestone	1.34
Slates, sandstones and thin limestones	30.00
Jenny Start Limestone	2.00 to 10.50
Slates	13.00 to 20.00

The slates of the sequence are silvery grey, calcareous and commonly silty, the sandstones generally grey, fine grained and seldom thicker than 0.3 m. Silty slates with *Chondrites* are much rarer than in the underlying Lester Slates-and-Sandstones. The top 5 m or so of the basal slates contain bands and lenses of detrital limestone up to 25 mm thick with crinoid debris. AW

The Jenny Start Limestone commonly has an oolitic texture, but the ooliths are in many cases deformed as a result of tectonism. At the eastern end of West Hagginton Beach the limestone takes on a strong brown colouration and is sideritised. At most exposures there is a varied coral fauna,

Plate 3 David's Stone Limestone in Sandy Bay, Combe Martin
Strata in the lower part of the David's Stone Limestone are folded into a syncline overturned to the north. (A 13023)

details of which are included in Table 4. Colonial rugose corals, many in the position of growth, are an important component of the fauna, including *Disphyllum aequiseptatum*, *Endophyllum* aff. *abditum*, *Thamnophyllum caespitosum* and *Xystriphyllum* aff. *quadrigeminum*. Tabulate corals also occur, including *Alveolites suborbicularis*, *Pachyfavosites polymorphus* and several forms of *Thamnopora*, of which the most common is *T. cervicornis*. Stromatoporoids are present, for example *Clathrodictyon sp.*, together with crinoids, probable cystoideans, brachiopods and gastropods. Holwill (1964b) noted the similarity of this coral fauna to that recorded by Webby (1964) from the Roadwater Limestone of west Somerset. Both are considered to be of Givetian age and to have accumulated in a shallow-water, subturbulent environment. Local absence of corals from the Jenny Start Limestone has been attributed to the action of bottom currents inhibiting coral growth (Holwill and others, 1969).

The strata between the Jenny Start Limestone and the Combe Martin Beach Limestone comprise calcareous slates with thin limestone bands in the lower part, some sandy strata in the middle part and well-cleaved grey slates in the upper part.

The distinctive Combe Martin Beach Limestone was described by Evans (1922) as highly ferruginous, with a considerable amount of pyrite, and associated with green calcareous shales. It has a characteristic coral fauna consisting predominantly of small solitary rugose corals, including *Barrandeophyllum?*, *Metriophyllum lituum* and *Syringaxon sp.*, associated with tabulate corals, fragmentary bryozoans, brachiopods, gastropods and crinoids; details are given in Table 4. Webby (1966b) inferred that deposition took place in deeper water than that of the Jenny Start Limestone.

Silvery grey calcareous slates with thin grey sandstones, locally cross-bedded, and thin dark grey crinoidal lenticular limestones, occur above the Combe Martin Beach Limestone and are exposed between Hele Bay and West Hagginton Beach and in the Sandy Bay area.

The David's Stone Limestone (Plate 3) near David's Stone [5703 4742] comprises two beds separated by slates. Corals are generally scarce (Table 4), except near the top of the lower bed and in two thin limestone bands above the upper bed. However, they are locally abundant at Broadstrand Beach [5318 4797], where Holwill (1964b) and Holwill and others (1969) regarded the large colonies of tabulate corals and stromatoporoids as forming a true reef. The corals are principally tabulate forms, including *Alveolites suborbicularis*, *Pachyfavosites* aff. *polymorphus*, *Thamnopora cervicornis* and *T. cronigera*. In contrast to the Jenny Start and Combe Martin Beach limestones, rugose corals are rare, being mainly represented by small specimens of *Syringaxon sp.* Stromatoporoids occur locally. Webby (1966b) related the presence of the corals to shallow subturbulent water conditions, and suggested correlation with the Holwell Limestone of west Somerset, which contains a varied coral fauna of Givetian aspect.   AW, DEW

Above the David's Stone Limestone is a poorly exposed sequence of grey, hard, silty slates, with some burrows and interbedded fine-grained sandstones and a few thin, dark grey but brown-weathering limestone bands. The top of the Combe Martin Slates is taken at about 30 m above the top of the David's Stone Limestone, a level determined by the prevalence of limestones below and thick sandstones above.

Inland, the Combe Martin Slates strike east-south-east. They give rise to a sharp angular landscape with abundant craggy scarps and pinnacles, in contrast to the rounded topography of the Lester Slates-and-Sandstones. Limestone brash occurs locally, but only in the vicinity of Combe Martin west and south-west of the River Umber have the limestones been extensively quarried. These quarries are now much obscured, but it is clear from the small exposures

Plate 4   Pinkworthy Pond
The pond spans the junction between Combe Martin Slates, exposed at water level in the middle distance, and Kentisbury Slates in the neighbourhood of the camera. (A 13010)

still visible that complicated tectonic structures of the kind seen on the coast are present.

The base of the Combe Martin Slates has been shifted south-eastwards by the Combe Martin Valley Fault and may be traced from Combe Martin via Buzzacott to Coulscott, where it is affected by faulting and probably folding. The top of the formation, however, is not displaced by the fault. East of Westleigh, the outcrop of the Combe Martin Slates narrows, and beyond Coulsworthy Lane [6204 4553] the formation is marked by only one topographical feature. The Westleigh area is also the easterly limit of the thicker limestone beds of the Combe Martin Slates, although thin dark grey crinoidal limestones are present farther east. AW

Combe Martin Slates crop out from just east of Blackmoor Gate [6469 4318] to the fault which runs south-east and east-south-east from Highley. Farther east, undivided slates between the Hangman Grits and the Kentisbury Slates (Plate 4)

extend from Chapman Barrows [695 435] to The Chains [732 425] and thence along the upper stretches of the Exe valley to pass beyond the district. The NNW-trending valley north and east of Hollacombe [6460 4378] follows a fault, and the top of the Combe Martin Slates shows a dextral displacement of 500 m.

East of the north-westerly fault near Highley, grey slates underlie a broad crop and extend north to the Hangman Grits. Beyond The Chains, these slates occupy a narrow crop and form the steep north-facing slopes of Long Chains Combe [743 422] and the upper Exe valley. The thickness of these eastern slates is difficult to estimate. Regional southerly dips, derived largely from the mapped junctions with sandier strata below and above, point to a thickness of about 200 m west of Highley but only some 120 m in the Exe valley. The Combe Martin Slates on the coast are 120 to 135 m thick (p. 29). EAE

Table 4 Ilfracombe Slates fossils

Total fauna identified	Jenny Start Limestone									Combe Martin Beach Limestone					David's Stone Limestone					
Site	A	B	C	D	E	F	G	H	I	J	K	L	M	N	O	P	Q	R	S	T
Coelenterata																				
HYDROZOA																				
1 *Clathrodictyon*	.	.	.	.	.	1	.	.	.	.	.	.	.	.	.	.	.	.	.	.
2 *Stromatopora*	.	.	2	.	.	.	.	.	.	.	.	.	.	.	.	.	.	.	.	.
3 stromatoporoids	.	.	.	3	.	.	.	.	.	.	.	.	.	.	.	.	.	.	.	3
ANTHOZOA																				
4 *Acanthophyllum*	.	.	.	.	.	.	.	.	4	.	.	.	.	.	.	.	.	.	.	.
5 *Alveolites suborbicularis*	5	5	5	.	.	5	5	5	.	.	.	5	.	.	5	.	.	.	.	.
6 *A. spp.*	.	6	6	.	.	6	6	.	.	.	.	.	.	.	.	.	.	.	.	?
7 *Aulocystis?*	.	.	.	.	.	.	7	.	.	.	.	.	.	.	.	.	.	.	.	.
8 *Aulopora*	.	.	.	.	.	.	.	.	.	.	.	.	.	.	.	.	.	.	.	8
9 *Barrandeophyllum?*	.	.	.	.	.	.	.	.	.	9	.	.	9	.	.	.	.	.	.	.
10 *Chaetetes multitabulatus*	.	.	.	.	.	.	.	.	.	.	.	.	.	.	.	.	.	.	.	10
11 *Coenites escharoides*	.	.	.	.	.	.	.	11	.	.	.	.	.	.	.	.	.	.	.	.
12 *C.?*	.	.	.	.	.	12	.	.	.	.	.	.	.	.	.	.	.	.	12	.
13 *Columnaria* aff. *junkerbergiana*	.	.	.	.	.	.	.	.	.	.	.	13	.	.	.	.	.	.	.	.
14 *Disphyllum aequiseptatum*	.	14	14	14	14	14	14	.	.	.	.	.	.	.	.	.	.	.	.	.
15 cf. *D. goldfussi*	.	.	.	.	.	.	15	.	.	.	.	.	.	.	.	.	.	.	.	.
16 *Endophyllum* aff. *abditum*	.	.	16	.	.	.	.	.	16	.	.	.	.	.	.	.	.	.	.	.
17 'Hexagonaria'	.	.	.	17	.	.	.	.	.	.	.	.	.	.	.	.	.	.	.	.
18 *Metriophyllum lituum*	.	.	.	.	.	.	.	.	.	.	.	.	.	18	.	.	.	.	.	.
19 *Pachyfavosites polymorphus*	19	19	19	?	.	.	.	.	.	.	.	.	.	.	.	?	aff.	aff.	.	19
20 cf. *Phillipsastrea devoniensis*	.	.	.	.	.	.	.	.	.	.	.	.	.	.	.	.	.	.	.	20
21 *Plasmophyllum (P.) secundum*	.	.	.	.	.	.	.	.	21	.	.	.	.	.	.	.	.	.	.	.
22 *P. (Mesophyllum)* cf. *sandhillense*	.	.	.	.	.	.	.	.	22	.	.	.	.	.	.	.	.	.	.	.
23 *Scruttonia*	.	.	.	.	.	.	.	.	23	.	.	.	.	.	.	.	.	.	.	.
24 *Stringophyllum*	.	.	.	.	.	.	.	.	24	.	.	.	.	.	.	.	.	.	.	.
25 *Syringaxon*	.	.	.	.	.	.	.	.	.	25	25	25	?	.	.	.	.	25	25	.
26 *Temnophyllum?*	.	.	.	.	26	.	.	.	.	.	.	.	.	.	.	.	.	.	.	.
27 *Thamnophyllum ?caespitosum*	.	27	27	cf.	27	27	27	27	.	.	.	.	.	.	.	.	.	.	.	.
28 *Thamnopora boloniensis*	.	.	28	.	.	.	.	.	.	.	.	.	.	.	.	.	.	.	.	.
29 *T. cervicornis*	.	29	.	?	.	29	29	29	.	.	.	.	.	.	29	29	29	.	29	29
30 *T. cronigera*	.	?	30	30	30	.	30	.	.	.	.	.	.	.	.	.	.	.	30	.
31 *T. polyforata*	.	.	.	.	.	cf.	31	.	.	.	.	.	.	.	.	.	.	.	.	.
32 *T.* cf. *tumefacta*	.	.	.	.	.	32	32	.	.	.	.	.	.	.	.	.	.	.	.	32
33 *T. spp.*	.	33	.	33	33	.	33	.	.	.	.	.	.	.	.	.	.	.	33	.
34 *Xystriphyllum* aff. *quadrigeminum*	34	.	.	.	.	.	.	.	.	.	.	.	.	.	.	.	.	.	.	.

Table 4 *continued*

Total fauna identified	Site	Jenny Start Limestone									Combe Martin Beach Limestone					David's Stone Limestone					
		A	B	C	D	E	F	G	H	I	J	K	L	M	N	O	P	Q	R	S	T
Bryozoa																					
35 fenestellid		.	.	.	.	.	.	.	.	.	?	?	.	?	.	.	35	.	.	35	.
36 rhabdomesid		.	.	.	.	.	.	.	.	.	.	.	.	.	.	.	.	.	.	.	.
37 indet.		.	.	.	.	.	.	.	.	.	.	.	.	.	.	.	.	37	.	.	.
Brachiopoda																					
ORTHIDA																					
38 *Schizophoria striatula*		.	.	.	.	.	.	.	.	.	.	.	.	.	.	.	.	.	.	.	.
39 *S.*		.	.	.	.	.	.	39	.	.	.	.	.	.	.	.	.	.	.	.	.
STROPHOMENIDA																					
40 *?Leptaena analogaeformis*		.	.	.	.	.	.	.	.	.	.	.	.	.	.	.	.	.	.	.	.
41 *L.*		.	.	.	.	.	.	.	.	.	.	.	.	.	.	.	.	.	.	.	.
42 indet.		.	.	.	.	.	.	.	.	.	42	.	.	.	.	.	.	.	.	.	.
RHYNCHONELLIDA																					
43 indet.		.	.	.	.	.	.	.	.	.	?	.	.	.	.	.	.	.	.	.	.
SPIRIFERIDA																					
44 cf. *Spinatrypa aspera*		.	.	.	.	.	.	.	.	.	.	.	.	.	.	.	.	.	.	.	.
45 *Mimatrypa desquamata*		.	.	.	.	.	.	.	.	.	.	.	.	.	.	.	.	.	.	.	.
46 atrypacean		.	.	.	.	.	.	.	.	.	?	.	.	.	.	.	.	.	.	.	.
47 *'Athyris'*		.	.	.	.	.	.	.	.	.	.	.	.	.	.	.	.	.	.	.	.
48 *Cyrtina heteroclita*		.	.	.	.	.	.	.	.	.	.	.	.	.	.	.	.	.	.	.	.
49 *Cyrtospirifer* cf. *verneuili*		.	.	.	.	.	.	.	.	.	.	.	.	.	.	.	.	.	.	.	.
TEREBRATULIDA																					
50 *'Rensselaeria'*		.	.	.	.	.	.	.	.	.	.	.	.	.	.	.	.	.	.	.	.
Mollusca																					
CEPHALOPODA																					
51 *'Orthoceras'*		.	.	.	.	.	.	.	.	.	?	.	.	.	.	.	.	.	.	.	.
52 ammonoid?		.	.	.	.	.	.	.	.	.	.	.	.	.	.	.	.	.	.	.	.
GASTROPODA																					
53 *Straparollus?*		.	.	.	.	.	.	.	.	.	.	.	.	.	.	.	.	.	.	.	.
Echinodermata																					
54 cystoidean		.	.	.	.	.	54	54	.	.	.	.	.	.	.	.	.	.	.	.	.
55 crinoid debris		55	.	.	.	.	55	.	.	.	55	55	.	.	.	55	55	.	55	55	.
Tentaculitoidea																					
56 *Dicricoconus?*		.	.	.	.	.	.	.	.	.	.	.	.	.	.	.	.	.	.	.	.
Trace fossils																					
57 *Chondrites*		.	.	.	.	.	.	.	.	.	.	.	.	.	.	.	.	.	.	.	.
58 indet.		.	.	.	.	.	.	.	.	.	.	.	.	.	.	.	.	.	.	.	.
		A	B	C	D	E	F	G	H	I	J	K	L	M	N	O	P	Q	R	S	T

The forms listed have been identified from material collected during the survey and from other collections. Occurrences at 17 sites (A–H, J–M, O–S) are shown. Relevant BGS specimen numbers, along with locality details are listed below. Coral forms recognised only in the Holwill Collection are listed in columns I, N, and T. Some of the taxa listed have not been recorded from any of the sites selected for inclusion in the table.

Localities and specimen numbers

A   Reef [5380 4804] 580 m at 243° from coastguard station, Rillage Point. DEA 432–452.
B   Cliff [5382 4804] 580 m at 242° from coastguard station, Rillage Point. DEA 58–94, 359–370.
C   Reef [5388 4809] 495 m at 245° from coastguard station, Rillage Point. DEA 684–721.

D   Cliff [5389 4804] 505 m at 238° from coastguard station, Rillage Point. DEA 98–133.
E   Fallen blocks [5402 4814] 340 m at 242° from coastguard station, Rillage Point. DEA 748–756.
F   Vertical faces [5429 4848] 170 m at 335° from coastguard station, Rillage Point. DEA 605–672.
G   Northern faces of old quarries at cliff top [5474 4835] 375 m at 085° from coastguard station, Rillage Point. DEA 757–849, 938–958.

H   Southern slopes of Jenny Start headland [5676 4752] 465 m at 005° from Home Barton, Berrynarbor. DEA 456–495.

I   Various localities; corals in Holwill Collection (Murchison Museum, Imperial College, University of London) not represented in BGS collections.

J   Reef [5405 4826] 275 m at 261° from coastguard station, Rillage Point. DEA 722–747.

K   Reef [5724 4739] 610 m at 058° from Home Barton, Berrynarbor. DEA 564–571.

L   Base of cliff [5726 4737] 620 m at 060° from Home Barton, Berrynarbor. DEA 559–563.

M   base of cliff [5731 4736] 660 m at 063° from Home Barton, Berrynarbor. DEA 555–556, 929–936.

N   As for column I.

O   Broadstrand Beach. Base of cliff [5321 4798] 700 m at 082° from Lantern Hill chapel, Ilfracombe. DEA 399–416, 917–928.

P   Hele Beach. Reef [5368 4790] 760 m at 238° from coastguard station, Rillage Point. DEA 417–426.

Q   Reef [5683 4740] 355 m at 018° from Home Barton, Berrynarbor. DEA 542–554, 596–604.

R   David's Hole. Base of cliff [5705 4738] 460 m at 046° from Home Barton, Berrynarbor. DEA 496–527.

S   Base of cliff [5706 4740] 475 m at 045° from Home Barton, Berrynarbor. DEA 572–595.

T   As for column I.

KENTISBURY SLATES

The Upper Ilfracombe Beds of Holwill (1964) and Holwill and others (1969) are the Kentisbury Slates of this account and consist of slates with sandstones, siltstones and subordinate thin limestones. The base of the formation lies about 30 m above the top of the David's Stone Limestone (p. 15), the top is the base of the overlying Morte Slates, and the Kentisbury Slates of the coastal section (Plate 1) are probably not more than about 150 m thick.

Generally the sequence comprises alternating groups of slate-rich and sandstone-rich units. The slate-rich units consist mainly of dark grey and silvery grey slates, commonly with lenticles (averaging 0.1 m long × 0.01 m thick) of silt or fine sand and beds of sandstone (up to 0.5 m thick) and limestone (up to 0.2 m). The sandstone-rich units contain massive, grey or pinkish grey, coarse-grained sandstones up to 2 m thick, slates, and some thinner and finer-grained sandstones up to 1 m thick which are locally cross-bedded. At least four sandstone-rich units are present near the coast, but the exact number is uncertain because of structural complications. Many of the slate units contain burrows of the *Chondrites* type, similar to those of the Lester Slates-and-Sandstones. The intermittent development of coarse detrital

Plate 5   Topography of the Kentisbury Slates near Ilfracombe
The sharp dip and scarp features to the right are in Kentisbury Slates; the more rounded feature to the left is formed by Morte Slates. (A 13033)

sands suggests some measure of instability in the source area to the north.

Evans' (1922; *in* Evans and Stubblefield, 1929) Woolscott Cleave Slates are here considered to be part of the Morte Slates. The top beds of the Kentisbury Slates, and thus of the Ilfracombe Slates, comprise silvery grey, smooth, calcareous slates with *Chondrites* mottles and current-bedded sandstones, and are exposed on the coast [5021 4714] south-west of Breakneck Point; above are the Morte Slates (p. 31). Inland, the Kentisbury Slates are poorly exposed, but their junction with the Morte Slates forms a good topographical feature and the sandstone-rich units produce ridges resembling dip-and-scarp topography (Plate 5). Thus at Torrs Park [5080 4718] ESE-trending ridges have steep north-facing slopes and gentle, southerly-inclined, dip slopes. Dip measurements on craggy north-facing outcrops confirm the inclination of the beds. Such features are less noticeable east-south-east of Yetland [5789 4500]. AW

The Kentisbury Slates form a broad outcrop from Mattocks Down to Simonsbath. Locally, they contain strong brown sandstone bands up to 3 m thick, although it is unlikely that sandstone comprises more than about 10 per cent of the total thickness, and they form the high ground of Mattocks Down, Kentisbury Down, Rowley Down, Challacombe Common, Chains Barrow, Dure Down and Prayway Head. Comer's Ground Quarry [6405 4243], south of Bridwick Farm and west of Wistlandpound, was opened for limestone, of which workable quantities are rare in the Kentisbury Slates. The horizon may be equivalent to that of the Leigh Barton Limestone of the Brendon Hills and the Quantocks.

The mapped boundaries of the Kentisbury Slates in the east of the district suggest a regional dip of about 10° to south or south-south-west and a thickness of some 330 m of strata, but some repetition by overfolding and faulting seems probable. Orchard (1979) referred conodonts from the Kentisbury Slates to the Upper *varcus* Subzone. EAE

Morte Slates

The Morte Slates underlie a band of country of fairly uniform width, 3000 to 3500 m, extending east-south-east from the north-west extremity of the district to pass beneath Berry Down, Arlington Court, Fullaford Down, Shoulsbarrow Common, Squallacombe, Burcombe and Great Woolcombe, south of Simonsbath. They are typically smooth silvery and greenish grey slates, with some silty slate and a little siltstone and sandstone, mainly in the lower part. Webby (1966b) envisaged deposition in shallow-water pro-delta and delta-platform environments, and the occurrence of thin sandstones in the lowest Morte Slates suggests a gradual change in the depositional conditions, with no major break. Quartz veins are abundant. Cleavage is commonly more in evidence than bedding, and points to the presence of northward-overturned folds whose axial planes dip southwards at between 60° and 70°. Development of cleavage has largely obliterated contained fossils, and only the spiriferaceans have proved of stratigraphic use.

The thickness of the formation is difficult to estimate. Rough calculations based on the relationship of local folding and apparent thickness, and the projection of this relationship over the whole width of outcrop, have suggested a max-

imum thickness of 1500 m (Edmonds and others, 1979). Regional dips taken from the mapped outcrops, ignoring the small-scale folding, point to a figure of around 600 m. It is not possible to be more precise. EAE, BJW, AW

On the coast, the topmost beds of the Kentisbury Slates (p. 30) give way to silty and sandy slates of the Morte Slates south-west of Brandy Cove Point [5021 4714] in a sequence of strongly faulted and overturned strata. The topographic feature associated with this change can be traced inland from Shield Tor [5261 4651] to Woolscott Cleave [5558 4517]. Slates at Woolscott Cleave were thought by Evans (1922) to be top Ilfracombe Slates, but are here referred to the lower Morte Slates. AW

Hicks (1896) provided the first descriptions of fossils from the Morte Slates, reporting various brachiopods and bivalves along with crustacean fragments and crinoid remains from the formation in the Ilfracombe area. He regarded the fossils as indicating a succession of Silurian and possibly Lower Devonian strata, older than any other rocks in north Devon. This supported his interpretation of the Ilfracombe Slates–Morte Slates junction as a major thrust, carrying the Morte Slates over younger rocks. Gregory (1897) disputed some of the determinations given by Hicks and regarded others as based on insufficient evidence; he thought the fossils suggestive of a Devonian rather than a Silurian age. Examination of Hicks's fossils, now in the Sedgwick Museum, justifies Gregory's more cautious approach to them. The material is largely badly preserved and distorted and, apart from *Lingula*, specimens are rarely determinable at generic level. Evans and Pocock (1912) cited the occurrence of '*Spirifer verneuili*' in the Morte Slates on the coast at Woolacombe as indicating an Upper Devonian age. Their material, which includes *Cyrtospirifer verneuili* sensu Vandercammen 1959, was not found strictly *in situ*, but evidence supporting its provenance from the formation has since been obtained (Edmonds and others, 1979) and the total evidence from the coast now suggests a Frasnian–Famennian age for the Morte Slates.

The fauna so far collected from the formation in north Devon includes *Lingula* and orthoid, spiriferacean and rhynchonellacean brachiopods, along with pectinacean bivalves and a probable *Modiomorpha*, possible crustacean fragments and crinoid columnals. Within the area covered by this memoir *Lingula mortensis* has been recovered from Lee; *L. mortensis*, a strophomenoid?, a rhynchonellacean?, *Pseudaviculopecten? mortensis* and crustacean? fragments from Mullacott; *L. mortensis* and an indeterminable pectinacean from Shelfin Wood; and crinoid columnals from Torrs Point. The fossiliferous outcrops of north Devon are too isolated and the faunas recovered too thin to provide any positive evidence to bear on Webby's opinion that the slates were the product of pro-delta and delta-platform environments. DEB

Pickwell Down Sandstones

The outcrop of the Pickwell Down Sandstones, which overlie the Morte Slates, extends from Fullabrook Down in the west to Long Holcombe in the east, then swings round a major syncline and complementary anticline with wavelengths of possibly 2000 or so metres and amplitudes of little less. Smaller repetitions in the west point to similar close to open folds of around one tenth this magnitude.

The formation consists of purple, red, brown and greenish grey sandstones commonly current-bedded and generally fine- to medium-grained. Subordinate red and grey shales are most plentiful towards the bottom and top of the formation. These rocks formed in the disturbed shallow waters of rivers, lakes, deltas, lagoons and seas; few fossils have been found in them, although Rogers (1919; 1926) obtained Upper Devonian fish fragments from the base of the formation on the coast and plant remains from exposures inland (Edmonds and others, 1979).

The base of the Pickwell Down Sandstones is marked by a sporadic but persistent band of tuff, whose outcrop has in many places been pitted for walling stone (Figure 2) and whose presence elsewhere is commonly reflected in the surface rubble. This rock, massive and strong where fresh, has been seen up to 8 m thick, and locally forms a distinct topographic feature. Early uncertainty about its nature, whether an intrusive felsite (Bonney, 1878) or a water-sorted volcanic ash with fish remains (Rogers, 1926), has given way to identification as a keratophyric tuff of contemporaneous volcanic origin.

The sandstones have withstood orogenic movements better than have adjoining slates. Dips are predominantly southerly or just west of south, locally 60° or more but generally no more than 45° and in places only 15° to 25°. Overturning is rare or absent. In the western part of their outcrop the rocks have been dug in many places, and distinctive surface brash of typically fine-grained purple sandstone abounds. Between Bratton Fleming and the River Bray, tuff debris is scattered along the basal outcrop of the formation, and the Pickwell Down Sandstones are well exposed in a number of old quarries. Exposures are less common eastwards towards Five Barrows Cross. Many fine-grained quartz-veined purple sandstone blocks occur on Fyldon Common and Long Holcombe. The high ground of North Molton Ridge and Barcombe Down is poorly exposed, but west of here the River Mole transects the outcrop between Heasley Mill and North Molton. Exposures are again uncommon towards the western end of this part of the outcrop, apart from small quarries in the Bray valley.

Extrapolation of local structural styles in the Bideford district suggested a maximum thickness of Pickwell Down Sandstones of about 1200 m (Edmonds and others, 1979). It seems no less in the west of the present district; indeed using the evidence of the largest exposure, Little Silver Quarry [549 403], it could be more. However, in the east, where the outcrop of the formation is narrower, the total thickness may be less than 1000 m.

Upcott Slates

The Upcott Slates are cream, buff, green, grey and purple slates and silty slates. Sandwiched between two stronger sandier formations the slates have accommodated more tectonic movement than have the immediately adjoining rocks. Cleavage and bedding commonly dip steeply to slightly west of south and there has been some overturning. A thickness of 250 m has been suggested for the coastal crop (Edmonds and others, 1979) and no more accurate estimate is possible inland.

The outcrop extends as a narrow band eastwards from Winsham [499 388] through Marwood and Shirwell. In the east the crop broadens around North Radworthy [752 341], in the core of the major syncline referred to above, and swings through North Heasley [732 332], Charles Bottom [683 313] and North Molton round the flanks of the adjoining anticline. The broadening is thought to be due largely to lower dip angles and some minor folding within the major syncline, although there may be some local thickening of the formation.

Baggy Sandstones

Some 450 m of interbedded fine-grained sandstones, siltstones and shales, with thicker buff fine and medium-grained feldspathic and micaceous sugary sandstones, crop out from Boode [501 380] to Stoke Rivers and Brayford, beyond which the crop may be traced around two major folds to pass south of North Molton and out of the district. These Baggy Sandstones generally form a marked rise to the south of the Upcott Slates. On the coast (Edmonds and others, 1979) the strata dip fairly uniformly southwards. Inland, dips of 30° to 60° between south-south-west and south-south-east predominate, with some vertical strata; a few northerly dips point to the presence of some close or open folds.

A broad crop of Baggy Sandstones extends east from Boode [501 380] to a fault trending south-east from Beara Charter Barton [5241 3838] to Whitehall Mill [5340 3746], beyond which the outcrop is much reduced in width. At 1 km east of Plaistow Quarry [568 373], the formation has been displaced about 1 km dextrally by a NW-trending fault. In the neighbourhood of Stoke Rivers the formation occupies a broader outcrop and extends west-south-westwards as three tongues in anticlinal cores.

There are good exposures of Baggy Sandstones on Mockham Down, but few in the fault-bounded block east of Brayford. However, those few suggest that much of the succession hereabouts is made up of fairly massive buff medium- to fine-grained feldspathic sandstone. South and east of there, the outcrop of the formation thins somewhat, and it is possible that the thickness of the Baggy Sandstones is reduced to about 300 m. Scattered exposures of brown and grey fine-grained sandstones with greenish grey shale partings occur in the syncline east of the River Bray, and exposures are uncommon in the anticlinal area around East Buckland. The tract of southerly-dipping Baggy Sandstones running west–east to the south of North Molton is mainly delineated by field fragments.

Goldring (1971) made a detailed study of the Baggy Sandstones on the coast, and concluded that they were mainly of marine origin but with some fresh-water and brackish-water sediments.   EAE, BJW

The fossils of the Baggy Sandstones are in general better preserved than those of older formations in north Devon, a factor which probably influenced the fuller descriptive treatment they have received in the literature. Both J. de C. Sowerby (in Sedgwick and Murchison, 1840) and Phillips (1841) dealt with elements of the fauna, but the main contribution to our knowledge of it lies in the painstaking descriptions provided by Whidborne (1896–1907). Other works include those of Arber and Goode (1915) on the plant remains, and Goldring (1962) on trace fossils from the formation. Partridge (1902) described forms of the phyllocarid crustacean *Echinocaris* from 'Sloley Quarry', probably referring to Plaistow Quarry, west of Sloley Barton. Faunal lists for the Baggy Sandstones were provided by Etheridge (1867) and Hall (1867), and Ussher (1879; 1881) noted the presence of the bivalve '*Cucullaea*' in some of the sandstones, which he called 'Cucullaea grits'. Goldring (1971) gave useful general notes on the faunas of the various included facies.

The Upper Devonian age of the strata has been generally accepted for over a century. Goldring (1970) referred them to the Famennian *Clymenia* Stufe but later (1971) transferred them, on conodont and spore evidence, to the topmost Devonian *Wocklumeria* Stufe. The conodont studies of Austin and others (1970) indicated that the uppermost Baggy Sandstones, along with the lowest Pilton Shales, are of very late Devonian age, younger than the uppermost *Wocklumeria* faunas of Germany but older than the highest Famennian faunas of Belgium.

The mapped junction between the Baggy Sandstones and the succeeding Pilton Shales is an arbitrary one, and separates locally variable deposits which resulted from the same general transition from terrestrial to open sea conditions. It is hardly surprising, therefore, that this junction has not been defined on palaeontological criteria, nor that faunal assemblages which appear to be typical of one occur locally in the other. Thus Goldring (1971) recorded productellid brachiopods, which characteristically occur in the Pilton Shales, from Baggy Sandstones on the coast; conversely, during the survey, faunas characterised by '*Cucullaea*', of typical Baggy Sandstones aspect, were collected in working quarries at Charles [e.g. 687 340], from strata mapped as Pilton Shales.

The varying sedimentary environments of the Baggy Sandstones are reflected in the faunal assemblages. Those deposits attributed by Goldring (1971) to distributaries and, possibly, fresh-water lakes yield plant debris, whereas those which he considered to have been formed in a sub-beach environment, along with certain marine deposits, are typified by trace fossils. Plants recorded from the present district include *Sphenopteridium rigidum* and *Knorria* from 'Sloly Quarry' (Arber and Goode, 1915). Invertebrate assemblages fall into three general types. One, characterised by '*Cucullaea*', in places accompanied by *Eoschizodus?* and *Palaeoneilo* along with bellerophontacean gastropods, is probably indicative of Goldring's Rough facies, which he thought to have been deposited near the shore and possibly in intertidal channel fills. A second is dominated by bivalves in variety, accompanied by numerous *Lingula* and scattered phyllocarid crustaceans and bellerophontacean gastropods, and is typical of a facies which Goldring regarded as restricted bay or lagoonal. During the survey, a fauna of this last type was collected from the northern end of the top south-east face of Plaistow Quarry [5688 3728]; the more common genera recorded, or close relatives, were similarly prominent in deposits of earlier Upper Devonian age encountered in the Willesden Borehole in south-east England, which have been interpreted (Butler, 1981) as of proximal inner shelf origin. The third assemblage, known from the coast, includes productellids and *Ptychopteria* and anticipates the more open marine conditions of the lower Pilton Shales.

Table 5 details faunas collected during the survey.   DEB

Table 5 Baggy Sandstones fossils

Total fauna identified	Locality	A	B	C	D	E	F	G	H
Plantae									
1 *Hostinella*		.	.	.	1	.	.	.	.
2 fragments indet.		2	.	.	.	.	2	.	2
Brachiopoda									
3 *Lingula circularis?*		.	.	.	3	.	.	.	.
4 *L.* cf. *squamiformis*		.	.	.	4	.	.	.	.
5 cf. *Ptychomaletoechia omaliusi*		.	.	.	.	.	.	5	.
Gastropoda									
6 bellerophontacean indet.		.	6	6	6	.	.	6	.
Bivalvia									
7 '*Cucullaea*' *unilateralis*		.	7	7	.	.	.	7	.
8 '*C.*'?		.	.	.	8	.	.	.	.
9 *Cypricardella?*		.	.	.	9	.	.	.	.
10 *Edmondia*		.	.	.	10	.	.	.	.
11 *Eoschizodus? deltoideus*		.	11	.	cf.	.	.	.	.
12 *Goniophora?*		.	.	.	.	.	.	12	.
13 *Leptodesma (L.) cimitum*		.	.	.	.	.	.	13	.
14 *Nuculoidea?*		.	.	.	14	.	.	14	.
15 *Palaeoneilo constricta*		.	.	.	15	.	.	.	.
16 *P. lirata*		.	.	.	16	.	.	?	.
17 *P. tensa*		.	.	.	17	.	.	.	.
18 *P.*		.	18	.	.	.	.	.	.
19 cf. *Paracyclas tenuis*		.	.	.	19	.	.	.	.
20 *P.*		.	.	.	.	.	.	20	.
21 *Prothyris contorta*		.	.	.	21	.	.	?	.
22 *Ptychopteria damnoniensis*		.	.	.	22	.	.	22	.
23 *Sanguinolites mimus*		.	.	.	.	.	.	23	.
24 *S. tiogensis*		.	.	.	24	.	.	.	.
25 cf. *Spathella munda*		.	.	.	25	.	.	.	.
26 cyrtodontid?		.	.	.	26	.	.	.	.
27 pterineid		.	27	.	.	.	.	.	.

Table 5 *continued*

Total fauna identified	Locality	A	B	C	D	E	F	G	H
Phyllocarida									
28 cf. *Echinocaris sloliensis*		.	.	.	28	.	.	.	.
29 *E.* cf. *socialis*		.	.	.	29	.	.	.	.
Crinoidea									
30 columnals indet.		.	.	.	.	.	.	30	.
Trace fossils									
31 burrows indet.		.	.	.	31	31	.	.	.
		A	B	C	D	E	F	G	H

The forms listed have been identified from material collected during the survey. Localities of occurrence (A–H) and relevant BGS specimen numbers are given below.

Localities and specimen numbers

A   Boode Quarry [506 378], 600 m at 315° from Ash Barton, Braunton; west face. DEA 2219–24.
B   Boode Quarry; north-east face. DEA 2225–40.
C   Plaistow Quarry, 500 m W of Sloley Barton, Shirwell; loose blocks [5684 3731], probably from adjacent face. DEA 2242–45.
D   Plaistow Quarry; northern end of top south-east face [5688 3728]. DEA 2246–2435.

E   Plaistow Quarry; top south-east face [5688 3725]. DEA 2436–2438.
F   Plaistow Quarry; south-west side of the main part of low level [5678 3721]. DEA 2439–2460.
G   Plaistow Quarry; southern part of highest south-east face [5684 3721]. DEA 2461–2517.
H   Quarry [7277 2918] 1400 m at 067° from South Lee, South Molton. DEA 2526–2529.

Pilton Shales

Shallow seas persisted over north Devon during the transition from Devonian to Carboniferous. Sedimentation was continuous throughout the deposition of the Baggy Sandstones, the Pilton Shales, and the overlying shales, cherts and limestones of the Codden Hill Chert. No sharp formation-boundary lines can be traced. The Pilton Shales are taken to begin above the highest of the typical fine- to medium-grained sugary sandstones of the Baggy Sandstones, and to end at the lowest dark grey to black locally siliceous shales of the Codden Hill Chert. They comprise shales and siltstones with limy and sandy bands and lenses. Brachiopods and bivalves of the lower part point to shallow waters near the shore or perhaps near a delta; trilobites and goniatites at higher levels indicate slightly deeper water.

Within the Barnstaple district the Pilton Shales, probably about 500 m thick, crop out from Heanton Punchardon through Barnstaple to West Buckland. Farther east, in the neighbourhood of Charles and East Buckland, they swing round major folds and continue eastwards in an outcrop narrowed to 1.5 km or less, with its southern margin roughly coincident with the line of the old Taunton to Barnstaple railway. Locally on both the northern and southern margins of the outcrop, the formation is deposited in upright open and close folds with amplitudes and wavelengths of a few hundreds of metres.

The shales, siltstones, thin sandstones and fossiliferous limestone lenses include thicker sandstones which commonly stand out as topographical features. In the west these harder bands appear to be commoner in the north of the outcrop than in the south, and hence possibly occur mainly in the lower part of the formation. They are rarely more than a few metres thick. Farther east, however, large quarries opened in strong thick-bedded and massive sandstones between Brayford and Charles (Plate 7) yield some of the hardest stone worked in north Devon.   EAE, BJW

The fossils of the Pilton Shales are generally in the form of moulds, some of which preserve considerable detail despite distortion. Early descriptions of them were included in the accounts of J. de C. Sowerby *in* Sedgwick and Murchison (1840, pls. 52–57) and Phillips (1841), after which followed a number of papers dealing with the fauna in more general terms, e.g. Hall, 1867. These works were largely superseded by the studies of Whidborne (1896–1907), whose careful descriptions provide the most extensive coverage of the fossils to date.

During the present century Reed first (1943) revised the brachiopods figured by Whidborne and later (1944) the trilobites. Goldring (1955) provided further information on the trilobites of the formation, along with valuable stratigraphic and locality details, and also briefly reviewed the goniatites. Later (1957) he gave a full account of the productellid brachiopods.

Paul (1937) correlated the Pilton Shales roughly with the Etroeungt Beds of the Continent. Goldring (1955) refined this; he divided the formation into three according to the distribution of various trilobites, referring the lowest faunal unit, Pilton A, to the Upper Devonian *Wocklumeria* Stufe and the remaining two to the Lower Carboniferous. Pilton A is characterised by *Phacops (Omegops) accipitrinus accipitrinus*; Pilton B, which he correlated with the *Gattendorfia* Stufe, has *Archegonus (Angustibole?)*, *A. (Phillibole)* and *Brachymetopus*, and Pilton C, probably of *Gattendorfia* or *Ammonellipsites* age, yields *Piltonia* and *Brachymetopus*. It is not certain whether Pilton C lies above or within Pilton B (Goldring, 1970). Some unfossiliferous shales were referred by Prentice (1960a) to a further group, Pilton D, lying above the faunal unit C. Meanwhile Goldring (1957) had split Pilton A into the ascending subdivisions A1, A2, A3, defined by the occurrences of species of productellid brachiopods. The studies of Austin and others (1970) suggested that the conodont faunas of the lower Pilton Shales, at least up to about 200 m above the base, were of very late Devonian age, younger than the youngest *Wocklumeria* Limestone conodont faunas of West Germany.

The transgressive phase, initiated in the Famennian, maintained its influence during Pilton Shales times, so that the lower strata were laid down in shallow seas which later gave way to more off-shore, deeper waters. Indications of near-shore conditions exist in large working quarries near Charles, 12 km east of Barnstaple, where decalcified bands yield bivalve-dominated assemblages (Table 6, column A) strongly reminiscent of those of the Rough and *Lingula* facies of the Baggy Sandstones. Elsewhere, the Devonian part of the formation, Pilton A, contains a varied fauna in which brachiopods are the most important element and which appears to reflect a more open, shallow sea. The brachiopods are diverse, though lingulids are lacking, and, where substantial faunas are retrieved, commonly include the orthoid *Aulacella interlineata, Crurithyris unguiculus* and other spiriferoids, rhynchonelloids and strophomenoids. Of this last order, chonetaceans and the productaceans *Hamlingella, Mesoplica* and *Whidbornella* are notable. Goldring's faunal subdivisions of Pilton A are based primarily on the productaceans, which are distributed as follows (Goldring, 1970):

Pilton A1, with *Whidbornella pauli, W. caperata* (occasional), *Hamlingella goergesi* and *Mesoplica praelonga*; without *H. piltonensis.*

Pilton A2, with *W. caperata, H. goergesi, M. praelonga* and *M. simplicior*; without *W. pauli* and *H. piltonensis.*

Pilton A3, with *W. caperata, H. piltonensis* and *M. simplicior*; without *H. goergesi.*

Bivalves remain an important faunal element with pterineids and pectinaceans prominent. Infaunal bivalves,

common in the nearer-shore deposits, occur only sporadically. Bryozoans and crinoid remains are widespread, and fragments of phacopid trilobites occur at several localities. Gastropods occur locally as minor faunal elements, and scattered ostracods and solitary corals are encountered.

Few assemblages from the Lower Carboniferous part of the formation are known in detail but brachiopods seem to remain the dominant faunal element and to consist largely of strophomenoids, spiriferoids and rhynchonelloids. The orthoid *Aulacella interlineata* has not been noted in these higher beds and may be confined to Pilton A. The genera *Hamlingella, Mesoplica* and *Whidbornella* are likewise lacking, other productaceans, including *"Productellina" fremingtonensis*, occurring in their stead. Bivalves are varied but of minor importance; they include several pectinacean forms but infaunal bivalves appear to be few. Solitary corals, gastropods and crinoid debris comprise the bulk of the remaining fauna. The phacopid trilobites of Pilton A are replaced by proetaceans which, though stratigraphically important as noted earlier, are not common. Rare goniatites, of the genera *Gattendorfia* and *Imitoceras*, occur.

Fossils identified from collections from the Pilton Shales of the present district are listed in Table 6.   DEB

In practice, Goldring's and Prentice's divisions cannot be mapped, neither can the Devonian–Carboniferous junction. The only mappable lines within the Pilton Shales are those delineating packets of sandstones, and even these commonly cannot be correlated with certainty across large faults. The thick sandstones in the valley of the River Bray south of Brayford crop out in both limbs of a major syncline (p. 38).

EAE, BJW

Table 6 Pilton Shales fossils

Total fauna identified	Site	Upper Devonian												?Upper Devonian					?Lower Carboniferous			Lower Carboniferous		
		A	B	C	D	E	F	G	H	I	J	K	L	M	N	O	P	Q	R	S	T	U	V	W
Plantae																								
1 wood fragment		.	.	.	.	1	.	.	.	.	.	.	.	.	.	.	.	.	.	.	.	.	.	.
Anthozoa																								
2 *Palaeacis?*		.	.	.	.	.	.	.	.	.	.	.	.	.	.	.	.	.	.	2	.	.	.	.
3 *Pleurodictyum*		.	3	.	.	.	.	.	.	.	.	.	.	.	.	.	.	.	.	.	.	.	.	.
4 *Syringaxon?*		.	.	.	.	.	.	.	.	.	.	.	.	.	.	.	.	.	.	4	.	.	.	.
5 solitary coral indet.		.	.	.	.	.	.	5	.	.	.	.	.	·5	.	?	.	.	5	.	5	5	5	5
Bryozoa																								
6 *Fenestella plebeia*		.	.	.	6	cf.	.	6	.	.	6	cf.	6	cf.	.	.	6	.	cf.	.	.	.	.	6	.
7 cf. *F. polyporata*		.	.	.	7	.	.	.	.	.	.	7	.	.	.	.	.	.	.	.	.	.	.	.
8 *F. umbrosa*		.	.	.	8	.	.	.	.	.	.	.	.	.	.	.	.	.	.	8	.	.	.	.
9 *F.*		.	9	9	.	.	.	9	.	9	.	.	.	.	9	.	.	9	.	.	.	.	?	.
10 *Penniretepora*		.	.	.	10	.	.	?	.	.	10	.	.	.	.	.	.	?	.	.	.	.	.	.
11 *Rhabdomeson*		.	11	.	.	.	.	.	.	.	.	.	.	.	.	.	.	.	.	.	.	.	.	.
Brachiopoda																								
LINGULACEA																								
12 *Lingula*		12	.	.	.	.	.	.	.	.	.	.	.	.	.	.	.	.	.	.	.	.	.	.
DISCINACEA																								
13 *Orbiculoidea?*		.	.	.	13	.	.	.	.	.	.	.	.	.	.	.	.	.	.	.	.	.	.	.
ENTELETACEA																								
14 *Aulacella interlineata*		.	14	14	14	cf.	.	.	14	.	14	?	.	14	14	cf.	14	cf.	.	.	.	.	.	.
15 cf. *Dalmanella whidbornei*		.	.	.	.	.	.	.	.	.	.	.	.	15	.	.	.	.	.	.	.	.	.	.
16 *Schizophoria resupinata*		.	.	.	.	.	.	.	.	.	.	.	.	.	.	.	.	.	.	.	.	16	.	.
17 *S.?*		.	.	.	.	.	.	.	.	.	.	.	.	.	.	.	.	.	.	.	.	.	.	17
18 dalmanellid indet.		.	.	.	.	.	.	.	.	.	.	.	.	.	.	.	.	.	.	.	.	18	.	.

Table 6   *continued*

	Upper Devonian												?Upper Devonian					?Lower Carboniferous			Lower Carboniferous		
Total fauna identified Site	A	B	C	D	E	F	G	H	I	J	K	L	M	N	O	P	Q	R	S	T	U	V	W
STROPHOMENACEA																							
19 'Leptaena'																		19					
20 *Leptagonia analoga*																					20	20	
21 *L.*																	21						
DAVIDSONIACEA																							
22 *Derbyia?*														22									
23 *Schuchertella?*																							23
24 cf. *Xystostrophia umbraculum*													24										
CHONETACEA																							
25 *Chonetes sauntonensis*			25	25				cf.	?	cf.	cf.	25		25		cf.							
26 *Rugosochonetes* cf. *hardrensis*																							26
27 cf. *R. laguessianus*																					27		
28 indet.				28									28		28		?	28	?				
PRODUCTACEA																							
29 cf. *Hamlingella goergesi*			29	29																			
30 *H. piltonensis*							30	30															
31 *Mesoplica praelonga*				31	31																		
32 cf. *M. simplicior*							32	?															
33 *?Ovatia spinulifera*																							33
34 "*Productellina*" *fremingtonensis*																							34
35 *Steinhagella steinhagei*							35																
36 *Whidbornella caperata*		?	36					36	36						?								
37 *W. pauli*		37	37	cf.																			
38 *W.*				38																			
39 indet.										?							?		39		39	39	
RHYNCHONELLACEA																							
40 cf. *Centrorhyncus letiensis*			?	40	40						40		40										
41 *C.?* *partridgiae*				41																			
42 cf. *Ptychomaletoechia omaliusi*	42																						
43 indet.		43				43	43	43		43	43				43			43			43		43
ATHYRIDACEA																							
44 cf. *Athyris concentrica*							44																
45 *A.?*															45								
46 *Cleiothyridina* cf. *fimbriata*									?				46										
47 *C.?*											47	47									47		
48 cf. *Composita struniensis*														?	48								
49 *C.*													49										
CYRTIACEA																							
50 *Crurithyris unguiculus*		cf.	50	50			50	50			50		cf.	50				50					
51 *C.*																			51				
SPIRIFERACEA																							
52 cf. *Cyrtospirifer grabaui*														52									
53 *C. verneuili* (sensu Vandercammen, 1959)				53							cf.												
54 *C.*							?								54								
55 *Septosyringothyris?*																					55		
56 cf. *Sphenospira julii*						56																	
57 '*Spirifer*' *mesomalus*				57																			
58 spinocyrtiid?																		58					
59 indet.		59		?						59	59		59										
SPIRIFERINACEA																							
60 *Punctospirifer*																					60		
61 *Spiriferellina*													?										61
RETICULARIACEA																							
62 *Reticularia?*																					62		
63 *Torynifer? microgemma*		cf.		63																			
64 elythid																					64		
STRINGOCEPHALACAEA																							
65 centronellid									65														

Table 6   *continued*

Total fauna identified	Site	Upper Devonian												?Upper Devonian					?Lower Carboniferous			Lower Carboniferous		
		A	B	C	D	E	F	G	H	I	J	K	L	M	N	O	P	Q	R	S	T	U	V	W
Gastropoda																								
66 *Bellerophon subglobatus*		66																						
67 *'Euomphalus'*																								67	
68 *'Loxonema'*									68																
69 *Murchisonia*									69																
70 *'M.'*																								70	
71 cf. *Naticopsis (N.) hallii*																		?						71	
72 *Platyceras?*																								72	
73 *'Pleurotomaria'*									73								73			73					
74 *Straparollus (Philoxene) vermis*																	74	74							
75 *S.?*									75																
76 bellerophontacean									76																
77 indet.				77							77														77
Cephalopoda																								
78 *Imitoceras*																								78	
79 cf. *'Orthoceras' speciosum*		79																						
80 *'O.'*									80																
Bivalvia																								
81 *Aviculopecten transversus*														81										
82 *A.*				82																				
83 cf. *'Cucullaea' depressa*		83																						
84 *'C.' unilateralis*		84																						
85 *'C.'*		85																						
86 *Cypricardella?*									86																
87 *Cypricardinia scalaris*		87																		87	87			
88 *C.*														88										88	
89 *Eoschizodus? deltoideus*		89																						
90 *E.* cf. *inflatus*		90																						
91 *E.?*					91																			
92 *Leptodesma (L.) cultellatum*												cf.					92							
93 *L.*		93		93																				
94 *Newellipecten?*																					94			
95 *Palaeoneilo* cf. *antiqua*		95																						
96 *P.* cf. *constricta*		96																						
97 *P. tensa*		97																						
98 *P.*									98											98				98	
99 *Paracyclas*																								99	
100 *Parallelodon*																				100				100	
101 *Polidevcia* cf. *sharmani*																								101	
102 *Prothyris* cf. *contorta*		102																						
103 *Pseudaviculopecten nexilis*				103	103													103						
104 *P.*									104																
105 *Pterinopecten?*																						105			
106 *Pterinopectinella austeni*																								106	
107 *Pteronites*														107										
108 *Ptychopteria (Actinopteria)* sp. ?nov.				108																				
109 cf. *P. (P.) damnoniensis*						109								109										
110 *P.*		110																						
111 *Sanguinolites mimus*		111																						
112 *S.?*		112																						
113 cf. *Spathella munda*		113																						
114 *Streblopteria lepis*																				114					
115 *S. piltonensis*				115								cf.								cf.					
116 pterineid?					116																			
117 pectinacean			117																					
118 modiomorphoid									118																
Trilobita																								
119 *Archegonus (Phillibole)*																								119	
120 *A. (Angustibole?) porteri*																								120	
121 *Phacops (Omegops) accipitrinus accipitrinus*			cf.		121	121	?	121	121			cf.												
122 *P.?*				122						122	122													
123 phacopid											123													
124 indet.														124									124		

Table 6 *continued*

Total fauna identified	Site	Upper Devonian												?Upper Devonian					?Lower Carboniferous			Lower Carboniferous		
		A	B	C	D	E	F	G	H	I	J	K	L	M	N	O	P	Q	R	S	T	U	V	W
Ostracoda																								
125 cf. *'Primitia' bovifrons*		.	.	.	.	.	.	.	.	.	.	.	125	.	.	.	.	.	.	.	.	.	.	.
126 indet.		.	.	.	126	.	.	.	.	.	.	.	.	.	.	.	.	.	.	.	.	.	.	.
Crinoidea																								
127 columnals		127	127	127	127	127	127	127	127	127	127	.	127	127	127	127	127	127	127	127	127	127	127	127
Asterozoa																								
128 *Drepanaster?*		.	128	.	.	.	.	.	.	.	.	.	.	.	.	.	.	.	.	.	.	.	.	.
Miscellaneous																								
129 *Conchicolites*		.	.	.	.	.	.	.	.	.	.	.	129	.	.	.	.	.	.	.	.	.	.	.
130 cornulitid?		.	.	.	.	.	.	.	.	.	.	.	.	.	.	.	.	130	.	.	.	.	.	.
		A	B	C	D	E	F	G	H	I	J	K	L	M	N	O	P	Q	R	S	T	U	V	W

The forms listed have been identified from material collected during the survey, augmented by a few well-localised museum specimens. Occurrences at 23 sites (A to W) are shown. Relevant BGS specimen numbers and stratigraphic assignments are given below. The table is intended as a supplement to the text and does not provide a complete list of fossils from the formation. Readers are referred to the literature, especially to Goldring (1955; 1957; 1970) and to Whidborne (1896–1907).

Localities and specimen numbers

A   Working quarries [687 340 to 691 329] near Charles, on west side of B3226 road. Similar to marine faunas of Baggy Sandstones. DEA 2755–2980.

B   Old quarry [560 351] 310 m at 340° from Westaway, West Pilton. Pilton A1. DEA 2530–2565.

C   Top Orchard Quarry [562 351] 320 m at 010° from Westaway, West Pilton. Pilton A1. DEA 2566–2618 and Geol. Soc. Coll. 6257–6258.

D   Exposures [7514 2735] along sides of track from Whitechapel Barton to Burwell, Bishop's Nympton. Pilton A, probably A1. DEA 3023–3129.

E   Old quarry [594 345] 220 m at 190° from Snapper, Goodleigh. Pilton A1 or A2. DEA 2619–2665.

F   Roadside exposure [5988 3570], 1720 m at 176° from Shirwell church. Pilton A1 or A2. DEA 2666–2688.

G   Trackside [5499 3446] 190 m at 174° from Bradiford House, West Pilton. Pilton A3. DEB 9340–9374.

H   Both sides of road [5503 3448] 190 m at 169° from Bradiford House, West Pilton. Pilton A3. DEB 9297–9339.

I   Quarry exposure [5785 3484] 1285 m at 298° from Yeotown, Goodleigh. Pilton A. DEB 9617–9631.

J   Floor of track [5986 3427] 830 m at 085° from Yeotown, Goodleigh. Pilton A. DEB 9582–9607.

K   Roadside [6251 3088] 810 m at 308° from Yeoland House, Swimbridge. Pilton A. DEA 2719–2754.

L   Crags [7206 2809] 820 m at 132° from South Lee, South Molton. Pilton A. DEA 2982–3022.

M   Small promontory [5108 3461] and reef immediately to its north, on Taw estuary around 1303 m at 138° from Heanton Punchardon church. Probably Pilton A. DEB 8328–8365.

N   Reef [5175 3384] 1330 m at 316° from Clampitt, Fremington. Probably Pilton A. DEB 9476–9490.

O   Lane section [5596 3441] around 370 m at 205° from Westaway, West Pilton. Probably Pilton A. DEB 9633–50.

P   Cutting [5610 3412] on north side of A39 road, 610 m at 184° from Westaway, West Pilton. Probably Pilton A. DEB 9243–9263.

Q   Cutting [5613 3414] on north side of A39 road, 605 m at 180° from Westaway, West Pilton. Probably Pilton A. DEB 9264–9278.

R   Section [4982 3572] on south side of lane, 457 m at 288° from Heanton Punchardon church. Fauna not diagnostic of any particular subdivision of Pilton Shales. Goldring (1955) recorded trilobites typical of Pilton C nearby [4980 3570]. DEB 8279–8327.

S   Railway cutting [5170 3339] 1085 m at 299° from Clampitt, Fremington. Lower Carboniferous? DEB 9375–9402.

T   Railway cutting, east side [5692 3216], 590 m at 167° from Sunny Bank railway station, Barnstaple. Lower Carboniferous? DEB 9287–9296.

U   Section [5395 3296] 110 m at 125° from Hollowcombe, Oakland Park South, Fremington. Lower Carboniferous. DEB 9543–9581.

V   Low cliff [5175 3349] 1100 m at 304° from Clampitt, Fremington. Pilton B. DEB 9403–9448.

W   Low cliff [5175 3350] 1100 m at 305° from Clampitt, Fremington. Lower Carboniferous. Goldring (1970) assigned the beds to Pilton B. DEB 9449–9475.

DETAILS

Lynton Slates

The lowest Lynton Slates exposed crop out west of Lynmouth below Yellow Stone [7062 4997] and Ruddy Ball [7157 5006]. In the easternmost coastal exposures of the Lynton Slates [7348 4954], adjacent to the major fault separating Lynton Slates from Hangman Grits (Foreland Grits) at Ninney's Well Bay, clasts of dark greenish grey silty and sandy shale and siltstone occur in dark grey slates and contrast strikingly with the host rock lithology. The clasts are up to 0.3 m long and 0.15 m wide, generally non-angular, and lithologically similar to rocks lower in the Lynton Slates sequence. According to Professor Simpson's notes, such a mélange also occurs close to the major fault at Myrtleberry Cleave [7417 4901] and north of the old limestone quarry of Waters Meet [7473 4872]. The clasts of the coastal exposures appear to die out laterally away from the fault within about 5 m, and yet persist vertically through an unknown thickness of not very intensely disturbed slates. This close association with a major fault suggests a causal relationship. The fault plane [7348 4954] is nearly vertical and sharp, with no sign of a structurally disturbed fault zone. Tectonic emplacement of the clasts seems most unlikely in such a structural context. Some quartz veining is associated with the fault, and drag of the south-dipping beds on the south side suggests that locally the fault plane dips to the south. North of the fault the strata are much more intensely disturbed.

Grey slates with sandstone bands, mapped as Lynton Slates, but with some sandstone bands akin to those of the Hangman Grits, crop out in the lower parts of cliffs from Ramsey Beach [6464 4938] to Woody Bay [6805 4888]. At Heddon's Mouth [6551 4961] transitional beds are present, but the upward passage from marine Lynton Slates into the continental sediments of the Hangman Grits is not well exposed because of faulting. The only good, readily accessible, exposures of the transition are in the neighbourhood of Great Burland Rocks [6650 4956], near Hollow Brook Combe. The junction is less easy of access at Woody Bay, as it is some way up a sheer cliff face, but in the eastern part of the bay [6825 4905] slates with sandstones pass upwards into massive quartzitic sandstones without argillaceous laminae; these sandstones may be seen near Woody Bay pier [6766 4900]. North-east of Slattenslade the slate–sandstone contact forms a strong feature [6860 4888] which may be followed inland.

An arenaceous division of the Lynton Slates may be traced from the west side of Crock Point [6858 4921] through the Lee Bay area to the Valley of Rocks and beyond. Fairly massive quartzitic sandstones with interbedded slates form prominent crags at the well-known landmarks of the Devil's Cheesewring [7051 4952] and Castle Rock [7046 4974], and also the craggy northern slopes of the Valley of Rocks known as Rugged Jack. These beds persist along the strike and have been quarried for building stone at Hollerday Hill [7140 4945]; similar strata crop out in the town of Lynton and eastwards along the East Lyn valley to the area of Watersmeet [7440 4866]. A palynological report by Dr B. Owens on a sample of Lynton Slates from a crag [7086 4992] west of Hollerday Hill noted carbonised spores with a distinctive ornamentation of processes with bifurcate terminations, probably referable to the genus *Hystricoporites* which ranges from Eifelian to lowermost Tournaisian.

More slaty beds crop out in the South Cleave area [7065 4945] and may be traced by sporadic, isolated exposures to the West Lyn valley between Barbrook and Lynton, and through Lyn Cleave [7278 4906] and Myrtleberry Cleave [7388 4860] to the southern slopes of the Brendon valley. At the last-named locality grey slates with sandstone bands form strong ESE-striking scarp features; the slates dip southwards, and between Cross Lane [7665 4780] and the Fellingscott area [7751 4770] they contain spiriferid brachiopods and fenestellid bryozoans.   AW

Hangman Grits

Hangman Point to Elwill Bay

Purple and grey sandstones and gritty sandstones dip at up to 40° southward on Little Hangman and locally [5851 4806] contain hematite. On the coast nearby [5884 4826] coarse-grained sandstones interbedded with pebble beds pass up into finer-grained sandstones. The pebbles are mainly of quartz, and similar rocks farther east [6018 4828], near Blackstone Point, contain large quartz pebbles together with clasts of green and purple slate.

Dr D. E. Butler has reported *Chondrites* and fragments of thick shells from loose blocks on the shore [582 479] west of Little Hangman, probably representing the Stringocephalus Beds; *Cypricardella?* and tentaculitoideans possibly of the genus *Volynites* from near the top of the cliff face [5930 4817], probably within the Sherrycombe Formation; and *Naticopsis?* and a myalinid? bivalve occurring as loose specimens in shallow pits [6134 4776].

Exposures [6248 4842] near a waterfall east of The Mare and Colt show purple-stained buff sandstones traversed by thin quartz veins and with slaty intercalations. Massive sandstone forming the small headland [6454 4920] between West Lymcove Beach and East Lymcove Beach is associated with thinner sandstones and slaty siltstones.

Great Hangman to Shilstone Hill

Small pits [6002 4808] just west of the summit of Great Hangman have been opened in purple gritty and pebbly sandstones with traces of purple slates. Old quarries farther east show: purple sandstone 1 m thick overlain by rubbly slabby sandstone with subordinate slates 1 m [6200 4763]; interbedded purple and buff sandstones and purple slates 3 m, beneath Head 2 m [6250 4814]; 8 m of purple and grey fine- to medium-grained sandstones with green slates and quartz veins [6415 4785]; and massive greenish grey argillaceous jointed sandstone 1 m, overlain by green slabby slaty sandstones and slates 1 m [6511 4657].

Roadside exposures north of Martinhoe show massive sandstone 6 m, overlain by flaggy siltstones [663 494]; and up to 5 m of cleaved grey silty sandstones and slates [668 493]. Beacon Down Quarry [666 459] exposes purple slates 5 m, beneath green coarse-grained sandstones 26 m. Green, angular-grained, irregularly jointed gritty sandstones crop out in Lane Head Quarry [6724 4542], and similar rocks are interbedded with green slates and slaty sandstones in a nearby road cutting [676 452].

Some 4 m of strata in a small pit [6877 4817] north-east of Croscombe Barton comprise fine-grained flaggy sandstones and siltstones, with the top siltstones affected by cryoturbation. Another small pit [7001 4791], west of Dean, shows 2 m of grey and greenish grey silty sandstones, siltstones and slates, the silty and sandy beds showing small-scale cross-bedding. Picket Gate Quarry [703 476] displays about 10 m of interbedded sandstones and slates.

On the eastern bank of the Farley Water, and near a faulted junction with Lynton Slates, a small quarry [7421 4731] shows over 5 m of greenish brown and greenish grey massive well-jointed sandstone with subordinate silty slates. Scob Hill rises to the east and bears several small pits in brown, grey and purple sandstones, commonly quartz-veined, with subordinate slates [7470 4684; 7500 4664; 7508 4688; 7540 4695; 7537 4653]. In Scobhill Quarry [7506 4654] 0.25 m of silvery grey and locally purple-stained slates in the core of a small anticline are overlain by 3 m of greenish and bluish grey fine-grained well-bedded jointed sandstone which is very micaceous in places.

Sillery Sands to Countisbury

The cliffs above Sillery Sands [740 498] contain much hard green fine-grained sandstone, commonly stained red and veined by quartz in the vicinity of small faults. Similar strata extend to

Foreland Point [755 512]. Interbedded siltstones are common locally [7474 5066], and at the northern end of the sands small-scale cross-lamination, cleavage and graded bedding show the strata to be right way up [7483 5059].

Exposures between Foreland Lighthouse and Caddow Combe are of purple and grey medium-grained sandstones and shaly sandstones, commonly well bedded but locally massive. Similar rocks, purple grey and green in colour and in places quartz-veined and irregularly jointed, crop out farther east. Dr D. E. Butler noted coalified plant fragments from scree [7569 5063].

Green sandstones, locally stained red, are common in the roadsides near Countisbury, but a small pit [7475 4990] to the north shows purple slates with subordinate sandstone. Similar small pits in irregularly jointed fine-grained sandstone lie a short distance to the east [7529 4967; 7526 4943].

Shallowford Common to Lanacombe

Sandstones, siltstones, slaty sandstones and slates near the top of the Hangman Grits are sporadically exposed in headwaters of the West Lyn River [711 438] on the south side of Shallowford Common and dip at about 40° south-west. Ruckham Combe [726 440 to 727 431] and associated stream channels show similar rocks dipping at 30° to 50° between south and west-south-west, with cleavage planes inclined in the same direction, commonly at slightly steeper angles. Slates and sandstones in Hoaroak Water [743 440 to 747 430] are overlain near the top of the formation [747 423] by thick-bedded massive grey buff fine-grained sandstones, locally micaceous, dipping at about 50° southwards.

Sandstones and slates in the upper reaches of Farley Water [755 440 to 756 431] dip at 35° to 65°/180° to 200°, and outcrops in the Hoccombe Water [777 434 to 785 434] show mainly grey and brown sandstones dipping at 30° to 40°/180° to 260°. Buff sandstone has been quarried [7839 4125] near the top of the Hangman Grits on the north side of the River Exe.   EAE

Ilfracombe Slates

WILD PEAR SLATES

Scattered exposures add little to the description in the general account (p. 13). Valley sections and hill outcrops in the Tattiscombe area [6295 4665] reveal silvery grey slates, with sandstones up to 0.15 m thick, in normal and inverted sequence; some inverted strata are disposed in small-scale zig-zag folds, and reverse faults are present. The small valley [6751 4400] immediately west of Highley [6785 4391] is cut in grey slates showing a cleavage dipping at 45° to 60°/195°.   AW, EAE

LESTER SLATES-AND-SANDSTONES

Burrow Nose to Lester Point

The outcrop of the Lester Slates-and-Sandstones comprises a western, discontinuous coastal part, west of Combe Martin Bay, and an eastern part which trends east-south-east across the district. Exposures at Rillage Point [5421 4867], Widmouth Head [5486 4859], The Warren [5562 4828] and near Small Mouth [5646 4786], although disturbed by overfolding and faulting, show slates rich in *Chondrites*, with sandstones and thin, dark grey, crinoidal limestones, typical of the Lester Slates-and-Sandstones. Easily accessible exposures in the area of Burrow Nose [5540 4851] (The Warren) are of overfolded slates and sandstones cut by reverse faults. The sandstones are brownish grey, up to 1.4 m thick, fine to medium-grained and cross-bedded. The Holey Limestone, 0.4 m thick and fossiliferous, crops out at Watermouth [5552 4826 and 5573 4827]. Lester Slates-and-Sandstones near Briery Cave [5593 4813] are commonly inverted, and intensely overfolded strata occur south-east of Small Mouth [5650 4772].

Near the type locality—Lester Point and Lester Cliff—the following succession was measured in the right-way-up limb of an overfolded anticline [5763 4747]:

	Thickness m
Siltstone and silty slate, grey, with lenses and bands of brown-weathering grey sandstone. Abundant *Chondrites* in places	seen 1.60
Sandstone, grey, medium-grained, rather lensoid	0.13
Slate, dark grey, silty, with abundant wispy partings or thin, elongate, lenticles of strongly iron-stained, paler grey siltstone and sandstone. Abundant *Chondrites*. A lenticular sandstone bed in the uppermost 0.2 m is up to 0.1 m thick. The sandstones are brownish grey and have laminations	0.90
Sandstone, grey, medium-grained, rather lensoid. Weathers brown. Locally forms two beds	0.8 to 0.18
Slate, rather dark grey with abundant wispy partings or thin, elongate lenticles of strongly iron-stained, paler grey siltstone. Abundant *Chondrites*. Quartz streaks parallel to bedding. A little disseminated pyrite is present, as are lenses of iron-stained siltstone up to 2 mm thick. A few calcareous lenses, laterally impersistent with crinoids in places	5.55
Limestone, dark grey, medium-grained, crinoidal and lenticular	0 to 0.03
Slate, dark grey, slightly silty, with thin paler grey siltstone lenticles. Strong iron staining along joints. Some *Chondrites*	1.38
Holey Limestone: Dark grey, medium-grained, crinoidal, and with thin muddy partings which separate about 5 beds of limestone. Corals and brachiopods. In places the limestone has a coarse honeycomb structure and is brown, particularly along joints or where weathered by the sea. Poorly and irregularly jointed	0.63
Limestone, dark grey, medium-grained. Comprises a nodular lower bed 0.04 to 0.12 m thick, with abundant crinoids and veined with calcite, and an upper bed, silty in part but containing a more calcareous band 0.02 m thick with shells	0.12 to 0.17
Slate and silty slate, grey to dark grey, with wispy lenses of paler grey siltstone. Some disseminated pyrite and abundant *Chondrites*	0.77
Sandstone, grey, fine- to medium-grained, finely laminated. Quartz veins parallel to bedding	0.11
Sandstone, grey, medium-grained, with iron-stained joints. A top bed, 0.19 m thick, is locally cross-bedded	0.46
Siltstone or silty sandstone, grey, fine- to medium-grained. In three beds with iron-stained joints	0.32
Slate, grey and dark grey, silty with bands of fine-grained, finely laminated sandstone from 0.07 m thick which increase in number and thickness towards top	1.23
Sandstones, grey, medium-grained, with fine current-laminations in places. Numerous siltstone partings	2.66
Slate, grey to dark grey, silty, with laterally persistent sandstone bands up to 0.06 m thick. Weathered brown along joints	0.85
Sandstone, brown-weathering, finely laminated	0.10
Slate, grey, silty, with some paler grey thin sandstone lenses	1.10
Sandstone, grey, mainly fine-grained and with fine laminations	0.32
Slate, dark grey, silty, with many thin wispy lenses of brown-stained, fine-grained sandstone. Some *Chondrites*	1.29

	Thickness m
Siltstones and sandstones, prominent beds capped by 0.6 m of hard, brownish grey medium-grained sandstone. The siltstones contain thin bands and long lenses of paler grey finer-grained sandstone. Traces of fine cross-bedding	0.27
Slates, grey to dark grey, silty, with laterally persistent sandstone bands up to 0.03 m thick	0.23
Sandstone, grey but weathering brown, fine- to medium-grained, up to 0.08 m thick, separated by argillaceous bands up to 0.01 m thick	0.25
Slates, grey to dark grey, silty, with interbedded fine-grained sandstones up to 0.05 m thick. Sandstones weather brown and increase in thickness towards the top	1.40
Slates, grey, silty	0.51
Slates, mainly grey but with a green tint, silty, with some thin bands of black slate	1.74
Sandstone, grey, finely laminated, with regular siltstone partings	0.20
Slates, mainly pale grey, silty, with many thin bands of silty sandstone. Two dark grey mudstone bands half-way up	1.22
Sandstone, grey, fine- to medium-grained, weathering brown along joints, finely laminated	0.08
Slates, grey, silty, with abundant lenticular partings and beds of fine-grained silty sandstone. Some *Chondrites*	0.67
Sandstones and silty sandstones, grey, fine-grained, weathering brown along joints, in 4 beds separated by partings of silvery grey silty mudstone with *Chondrites*	0.61
Sandstones, grey, medium-grained, locally current-bedded, interbedded with grey silty mudstone with *Chondrites*	1.32
Sandstone, pale grey, coarse-grained and gritty, weathering to violet and pale purple. Lenticular lower bed up to 0.35 m thick	0.70 to 0.90
Slates, silvery grey, silty, with wisps and lenticles of fine sandstone. Abundant *Chondrites*	0.20 to 0.45
Sandstone, grey, medium-grained, with fine cross-lamination near middle	0.48
Sandstones and siltstones, grey, rather wispy bedded with mudstone wisps	0.90
Siltstones, dark grey, and sandstones, pale grey. Well-bedded	0.31
Sandstones, grey, medium-grained, with many horizons of dark grey siltstone in the top 0.05 m and the bottom 0.08 m	0.29
Sandstones and silty slates, interbedded, grey, the sandstones medium-grained and commonly cross-bedded	2.76
Sandstones, grey, medium-grained, cross-bedded, with silty partings up to 0.01 m thick. Dappled appearance where weathered. Disseminated galena specks	seen 1.68

AW

Hollacombe to Highley

Silty slates with thin sandstones are exposed in the valley [6490 4398] north-east of Hollacombe. In the old railway cutting [6620 4406] at Rowley Cross, brown sandstones and slates dip at 30° to 35°/180° to 200°. Upstream (southwards) from the Wild Pear Slates in the small valley west of Highley [6785 4391] occur a few scattered exposures of slates with slaty sandstone; fine-grained sandstone is common in the surface brash thereabouts.   EAE

COMBE MARTIN SLATES

Ilfracombe to Combe Martin Bay

At Jenny Start [5676 4752], in Sandy Bay, the Jenny Start Limestone comprises a lower 4.6 m of massive, poorly bedded limestone passing up into 4.6 to 5.5 m of thinly bedded, rather earthy, argillaceous limestone with an abundant coral fauna in which *Endophyllum* aff. *abditum* and *Thamnopora cervicornis* are particularly common (Table 4, column H). The most westerly coastal exposures of the limestone occur in Hele Bay [5380 4806; 5390 4811], where the beds lie in an easterly-plunging faulted anticline; *Alveolites spp., Disphyllum aequiseptatum, E.* aff. *abditum, Pachyfavosites polymorphus, Thamnophyllum caespitosum* and *Thamnopora spp.* are well represented (Table 4, columns A–D). The outcrop is traceable inland to the edge of the golf course [5420 4813]. Coralliferous Jenny Start Limestone crops out on the foreshore and in the cliff at the east end of West Hagginton Beach [5420 4845; 5425 4840], and can be followed eastwards along the strike through old workings south of Rillage Point to Samson's Bay [5457 4837] and south of Widmouth Head [5471 4834]. Vertical faces at the northern end of Rillage Point [5429 4848] and old quarries [5474 4835] south of Widmouth Head have yielded a varied fauna (Table 4, columns F, G) in which *D. aequiseptatum, T. caespitosum* (including a sub-cerioid form) and *Thamnopora spp.* are particularly common. From the old quarries, Dr D. E. Butler has identified the brachiopod *Schizophoria sp.* Farther east, the limestone is exposed high in the cliffs at Golden Cove [5659 4762]. West of Jenny Start, it is visible at extreme low tide in Combe Martin Bay [5740 4743].

The Combe Martin Beach Limestone is well exposed near Hele Bay [5375 4803] and at West Hagginton Beach [5402 4823], where it is intensely faulted and folded, and contains the small solitary corals *Barrandeophyllum?* and *Syringaxon sp.* Bryozoans, brachiopod fragments (including *Cyrtospirifer verneuili* according to Holwill and others, 1969) and orthocone fragments are also present (Table 4, column J). Further outcrops occur between Sandy Bay (The Parlour) [5721 4738] and Combe Martin Beach [5766 4728], but only on the west side of the Combe Martin Fault. Small, solitary, rugose corals are again well represented and a single example of the compound rugose form *Columnaria* aff. *junkerbergiana* was also found (Table 4, columns K–M). The Combe Martin Beach Limestone at Newberry Beach [5740 4732] is typical of the coastal exposures of the limestone, and the sequence is as follows:

	Thickness m
Limestone, dark grey, fine- to medium-grained, lenticular, fossiliferous and iron-stained	0.02 to 0.10
Slate, green, silty	0 to 0.05
Limestone, purplish brown weathered but grey-hearted, fine-grained, with crinoids and brachiopods. Scattered wisps of green silty mudstone. Orange-brown iron-stained irregular joints	0.55
Slate, green, silty, with worm burrows	0 to 0.03
Limestone, grey-hearted, iron-stained, rather coarse-grained, with some ferruginous shells	0.22
Slate, green, silty, lenticular, rather wispy, with some worm burrows	0 to 0.03
Limestone, grey-hearted, iron-stained, with a thin band of purple sandstone in the lowest 0.05 m. Poorly defined and weathered to a honeycomb structure	about 0.12
Slate ranging to siltstone, greenish grey to green, with worm burrows	0.09
Limestone, grey to dark grey, rather fine-grained, with crinoids and brachiopods. Weathered to pinkish grey in places	0.23

The green (?chloritic) slate occurs not only as discrete beds and lenses but also as wisps and pellets within the limestones. It was from outcrops of this limestone in the vicinity of Newberry Beach that Holwill (1964a, pp. 116–117) collected and described *Metriophyllum lituum.*

The David's Stone Limestone (Plate 3) crops out near David's Stone [5703 4742] in Sandy Bay, at the foot of the path to the beach. It comprises a lower bed of massive, dark grey, crinoidal limestone about 3.5 m thick, a middle bed of dark grey, calcareous slates about 2 m thick, and an upper bed of massive, dark grey limestone about 3.5 m thick. The top 0.3 to 0.5 m of the lower bed is very fossiliferous impure limestone. From the top 1 m of the upper limestone numerous examples of *Pachyfavosites* aff. *polymorphus* and *Thamnopora spp.* were collected, associated with *Syringaxon sp.*, crinoid debris and bryozoans (Table 4, column R). Above the David's Stone Limestone, and separated from it by 0.6 to 1.8 m of calcareous shales, are two bands of detrital limestone 0.08 to 0.15 m thick, with crinoid columnals, bryozoans and the corals *Syringaxon sp.* (common) and *P. polymorphus.*

The fauna (Table 4, column Q), lithology and stratigraphy of Evans's (1922) Red Limestone [5682 4741] are like those of the David's Stone Limestone, and undoubtedly the two are equivalent. The red colour is most likely due to the proximity of fractures. The tectonics are complex, but it seems probable that the isolated stack of red limestone is the higher limestone of the Davids Stone Limestone, folded upon itself with normal limb resting on inverted limb. The key to this interpretation is the recognition of the thin 'wriggly' limestone that occurs above the David's Stone Limestone. In the cliff [5676 4742] to the west, and separated from the stack by a fault, exposures of the Red Limestone show the distinctive colour only low in the section. They reveal a repetition of the higher limestone of the David's Stone Limestone, with superincumbent 'wriggly' limestone. The exposures are separated by a thrust fault and the repeated bed is in normal sequence.

The higher bed of the David's Stone Limestone crops out in inverted sequence at David's Hole [5693 4735], a NW-trending cleft along a fault plane. Other exposures of the bed occur at Broadstrand Beach [5318 4797], where *Alveolites suborbicularis, Thamnopora cervicornis,* rugose coral fragments and crinoid debris are present (Table 4, column O), and in Hele Bay [5369 4793] where *A. suborbicularis, P. polymorphus?, T. cervicornis* and fenestellid bryozoans occur (Table 4, column P). At both places the overlying calcareous slates and thin 'wriggly' limestones are visible. AW, DEW

Hodges to Coulscott

Thin dark grey crinoidal limestones exposed in Rectory Quarries [5842 4559] and Eastacott Quarries [5865 4561] are probably those just above the David's Stone Limestone. They are tectonically thickened by overfolding and overthrusting.

The junction between the Lester Slates-and-Sandstones and Combe Martin Slates is well exposed in a lane [6042 4615] east of Combe Martin where southerly dipping sandstones and silty slates with *Chondrites* give way southwards to silvery grey calcareous slates; the junction coincides with a topographic feature. AW

Hollacombe to Warren Farm

East of Hollacombe, grey slates exhibit cleavage inclined at 35° to 65° southerly. Sandstone occurs as fragments, but rarely if ever as outcrop. Old quarries on the northern slopes of Rowley Down may have been opened in slates and sandstones but they are now obscured.

An exposure [6839 4397] east of Highley showing a 1.2-m sandstone band in slates yielded a bedding dip of 35°/220°. Silty slates are common to the east, as in the northern end [7235 4238] of Pinkworthy Pond (Plate 4). Exposures in Long Chains Combe, just above the Hangman Grits, show silty slates with some fine-grained

slaty sandstone [7374 4228]. Rocks farther east, in a similar stratigraphical position, comprise silty slates with thin purple-spotted sandstones [7470 4223]. South-dipping silty slates with wisps and lenses of siltstone and sandstone crop out in clefts and crags on the south side of the River Exe [768 411; 7695 4105; 7673 4089]. EAE

KENTISBURY SLATES

Brandy Cove Point to Ilfracombe

The lowest 15 to 20 m of the Kentisbury Slates, comprising slates with sandstones, crop out in the area of Rapparree Cove [5283 4775] and the Fisherman's Chapel [5251 4790], and contain a 3-m thickness of distinctive purplish brown coarse-grained sandstone with thin slate intercalations. They are overlain by about 20 m of slates with a few thin sandstones, succeeded by 25 to 30 m of slates with thicker sandstones well exposed at Capstone Point [5192 4794] (Plate 1) and Wildersmouth Beach [5188 4786]. These latter strata contain beds of massive, purplish brown medium- to coarse-grained sandstone up to 2.0 m thick interbedded with grey, *Chondrites*-rich slates, and are much folded and faulted and well cleaved. Two higher sandstone-rich successions crop out at Tunnel Beach [5145 4781] and west-south-westwards along the coast to Brandy Cove Point [5047 4756]; they are 5 to 10 m thick and are repeated in many places by folding and faulting. AW

Brinscott to Simonsbath

Grey and greenish grey slates and silty slates are exposed around Brinscott [5866 4380], where the higher ground carries debris of purple-stained sandstone. Greatgate Quarry [6130 4250], near Patchole, contains grey silty slates with grey and purple fine-grained sandstone beds up to 0.3 m thick. Thin quartz veins contain some euhedral quartz crystals, and traces of current bedding within the sandstones show the strata to be right way up. In another quarry [6227 4424], north of Kentisbury, a 1.5-m bed of sandstone has been dug. Kentisbury Down is littered with blocks and boulders of buff and brown, fine-grained and medium-grained, massive sandstone. Many small pits are grassed over, but an old quarry [6435 4373] west of Hollacombe shows thick-bedded fine-grained sandstones up to 3 m thick, dips of 50° to 65° southerly and evidence of overfolding (p. 64).

Exposures in and around Comer's Ground Quarry [6405 4243] show slates and silty slates with some siltstone and silty sandstone, limy lenses, thin limestones, calcite veinlets and dips of 45° to 75° southerly. A well-preserved old limekiln [6398 4255] stands alongside the access lane and there seems little doubt that the workings were opened for limestone. No workable quantity of limestone remains visible, and the quarry is flooded; probably therefore the occurrence was lenticular and has been worked to its full lateral extent and either its full depth or the greatest depth practicable.

A boulder in the disused Indown Quarry [6576 4202] is of pale buff coarse-grained feldspathic sandstone containing slate fragments up to 7 mm across. Surface brash and traces in overgrown quarries on Rowley Down point to lithologies similar to those of Kentisbury Down (p. 30), and the same is true of Higher Down [673 422] and Challacombe Common. Exposures on the steep slopes between Challacombe reservoir and Swincombe Rocks [696 425] to the north comprise mainly slates and silty slates, and lithological distinctions between these rocks and the underlying Combe Martin Slates are vague. Similar rocks have been dug from Southground Quarries [6929 4127], where they contain a few thin limy lenses. Moderate to steep southerly dips prevail, although northerly dips occur locally, as on the north side of North Regis Common. Much sandstone rubble is scattered hereabouts, and also on Pinkworthy and Goat Hill, but slates with thin siltstones and sandstones have been worked at North Regis Quarries [7103 4126],

Short Combe Rocks [7250 4076] and Driver Quarry [7364 4027]. A good deal of slaty sandstone is exposed in The Chains Valley [747 418] and Tangs Bottom [751 404], and sandstone debris is common farther east on the high ground immediately north of Simonsbath.   EAE

Morte Slates

Torrs Park to the Sterridge valley

The lowest Morte Slates are exposed on the coast [4972 4720] as bluish grey or silvery grey slates, commonly silty. Interbedded thin sandstone bands display intense folding. The slates are well cleaved and the strata are overfolded and faulted. Lowest Morte Slates in craggy outcrops near Cairn Top [5155 4627] are overfolded and cut by quartz veins. North of Score [5200 4604] cleavage dip swings locally from south-south-west to east-south-east.

Excellent road sections [5150 4521 to 5172 4561] near Mullacott Quarries lie close to exposures at Mullacott Quarries [5171 4544] that were examined by Hicks (1896). Bluish and greenish grey silty slates are cut by quartz veins. The slates are disturbed and well cleaved, and both host rock and quartz veins are stained pink. Joints, cleavage and faults are stained pink and brown, and in places the cleavage is wavy with kink bands.

The feature associated with the base of the Morte Slates is readily traceable east-south-east from the Shield Tor area [5261 4651] to Woolscott Cleave [5558 4517] in the Sterridge valley. Exposures at Woolscott Cleave are of bluish grey silty slates with a strong southerly-inclined cleavage.   AW

Trimstone to Churchill

An old quarry [5000 4348] at Trimstone shows a 6-m face of slates and silty slates whose cleavage dips at 60° to 75°/185° to 195°. Exposures alongside the old railway to the south [5019 4231 to 5035 4225] show the youngest rocks of the formation to comprise slates with rust-spotted sandstones. Slates forming craggy outcrops on the slopes of a small valley [519 425 to 520 429] north of West Down contain many thick quartz veins. Abundant exposure of slates and silty slates characterises the valley of the River Caen between Lower Aylescott [5255 4164] and West Stowford Barton [535 428].

Berry Down is marked by several small disused pits [5622 4332; 5625 4286; 5709 4326; 5760 4336] whose exposures and rubble show that the Morte Slates of this high ground contain beds of greyish green fine-grained sandstone dipping at 60° to 85°/165° to 170°. An east–west line of small diggings [570 427] probably followed sandstone bands along strike, and an old pit [5767 4269] to the east shows slates and silty slates with fine-grained sandstone, vertical and trending 090°. An old pit [5697 4170] near Indicott contains slates and siltstones with sandy lenses vertical or dipping steeply southwards.

Farther east, fine-grained sandstones, siltstones and slates dip at 55° to 90°/185° in a small pit [5875 4245] and steeply northwards and southwards in the neighbourhood of Ford [5930 4235]. The higher strata of the formation, exposed to the south around Clifton [598 414] and Churchill [595 408], contain fewer sandy bands.

Arlington to Shoulsbarrow

Good exposures of Morte Slates occur at Rock Cottage [6029 4102] and in Parsonage Quarry [6065 4118]. Arlington Court [611 405] is underlain mainly by slates but Arlington Beccott [618 418] is located on lower Morte Slates, and surrounding exposures show the presence of thin-bedded and slaty sandstones. Lock Lane Quarry [6279 4143] to the east shows slates and silty slates with silty lenses and a little purple and brown fine-grained sandstone dipping at 85°/175°. A lane leading north off the A39 road east of Kentisbury Ford crosses a small east–west valley [6232 4276] and exposes slates

and silty slates; these are probably basal Morte Slates lying in a syncline, with the axis of a complementary anticline running approximately along the line of the A39 road.

Immediately south of Wistlandpound Reservoir slates and silty slates contain subordinate fine-grained fissile and slaty sandstones, mainly showing steep southerly dips in bedding and cleavage. These give way southwards to slates with scattered sandstone bands in the triangle defined by Twitchen [6405 4033], South Thorne [6459 4079] and Hunnacott [6479 4026]. Small mounds [6393 4011] near Twitchen show slates and silty slates with black earthy films, and veinlets up to 10 mm across, mainly of manganese oxide with some iron oxide and traces of copper. Slates near the top of the formation are extensively exposed on the steep slopes of Tidicombe Wood [639 393] and along the east–west stretch of old railway track just north of Bratton Fleming station [6413 3839 to 6468 3858]. Bedding and cleavage dip steeply south or are nearly vertical, and the rocks are more uniformly argillaceous and less silty than those to the north.

Slates and silty slates are extensively exposed in a small valley [6539 4100 to 6502 4037] running south-south-west from Stowford; they contain sandy streaks, and a 300-m east–west line of old quarries to the east [659 404], now largely filled, shows slates and silty slates with a little purple sandstone dipping at 75° to 90°/175° to 180°. Another old quarry [6798 4107] shows slates and silty slates with scattered sandy bands. Exposures in the north–south valley west of Barton Town [6800 4056] are almost entirely of slates and silty slates. Similar rocks in the River Bray south of Barton Town show a few sandy streaks. Sandy streaks are evident in silty slates alongside the River Bray upstream of Challacombe Mill [6810 4038]. The floor of Rocky Lane [6940 4002], near Shoulsbarrow, is a bare worn pavement of silty slates with a little fine-grained sandstone. Farther east, silty slates exposed in Weirs Combe [7005 4006] contain some thin current-bedded fine-grained sandstones striking 080° to 090° and vertical or steeply inclined to the south   EAE

Bratton Down to Great Woolcombe

Some 1.5 m of greenish grey slate dipping at 80°/170° are seen in a roadside exposure [6527 3931], and similar slates with silty bands dip at 70°/180° nearby [6528 3927]. About 500 m to the south, 1 m of grey silty slate sips at 70°/165°. Most of the old quarries on Bratton Down are now filled in and cultivated, but in one [6626 3917] 3.5 m of grey rubbly slate are overlain by 3 m of hard grey slate, dipping at 80°/175°. About 5 m of rather soft grey shale crop out [6727 3958] near Kipscombe, and farther south along the Bray valley sporadic exposures occur on Leworthy Hill, including 5 m of dark grey slate dipping at 80°/350° [6727 3833] and 3.5 m of variegated slates dipping at 70°/150° [6722 3825]. Farther east, 5 m of grey shaly slate were exposed during road improvements at Fullaford Hill [6832 3796], dipping at 70°/355°. There are many old pits and apparent trial holes at Wallover Down; among the larger exposures are 8 m of green slate [6889 3941], 6 m of pale green slate [6895 3925], 3 m of brownish grey slate [6941 3910] and 4 m of green slate with thin sandstones [6974 3929] in Goat Combe.

Fragments of reddened slates with quartz veinstone debris are common on Shoulsbarrow Common. The disused Castle Quarries on Castle Common [706 388] reveal largely overgrown exposures mainly of green slate, with some silty and sandy bands, dipping at 50° to 80°/355°, and similar rocks were exposed in a network of drainage ditches in the fields to the immediate south [705 386]. Sporadic exposures of mainly green slates occur in the combes leading south-west from the high ground of Henthitchen and Bray Common, but exposures are rare on the high ground owing to a thick head mantle and a thin discontinuous peat cover.

Most of the exposures between Ricksy Ball and Great Woolcombe are in the steep-sided valleys of northward-flowing tributaries to the River Barle. In valleyside crags [7404 3825 to 7406 3850] greyish green slates with thin sandy bands and thin sandstone beds are exposed; similar green slates with thin sandstones crop out in the valley to the east [7448 3842 to 7846 3860], and also in an old quarry [7470 3890] on the east bank of the Barle. A good section is exposed in the stream flowing northwards to join the Barle at Cornham Ford. Crags just north of the ford [749 387] (Plate 6) reveal green slates with rare sandy bands, but the streamside crags [7490 3855 to 7495 3828] are composed of green slates with many sandy beds up to 0.1 m thick. Near the source of the stream, red iron-stained shales are exposed [7516 3808] dipping at 50°/190°, and 100 m to the south-east [7522 3798] black slates dip at 70°/200°. Drybridge Combe cuts deeply into the Morte Slates, and afford a good section [7585 3799 to 7588 3866], mostly in green slates with thin sandstones, with red-streaked green slates exposed at one place [7583 3836]. The Brayford–Simonsbath road traverses the head and eastern flank of the combe, and roadside quarries reveal good exposures; 4 m of grey slates are exposed in one quarry [7592 3791] and 8.5 m of green slates with thin sandy beds in another [7606 3806]. A third quarry [7608 3827] contains 4.25 m of highly cleaved green slate with thin hard sandy ribs. A large quarry [761 385] has been excavated in 15 m of hard green slate with thin sandstones up to 0.1 m in thickness; about 1.5 m of black slate crop out at one place [7624 3863], and 10 m of greenish grey slates seen at another site [7627 3869] lie near the base of the formation.

Farther east, greyish green slates with scattered thin sandstones and sandy bands crop out in crags bordering NE-flowing streams [7700 3825 to 7724 3831; 7762 3802 to 7800 3808]. Greyish green slates, locally silty, are exposed in crags on the sides of the Barle valley [788 376] at the eastern margin of the district. Exposures are uncommon on the high ground south and west of the Barle, but green slates with thin sandstones can be found [765 372] in the banks of a tributary of Sherdon Water.   BJW

Pickwell Down Sandstones

Stoneyard Wood to the Bradiford Water

Some 3.5 m of weathered tuff at the base of the Pickwell Down Sandstones crop out [5041 4221] beside the old railway west of West Down, at the western edge of the district. A NW–SE fault runs along the valley to the east, and exposures of tuff, now obliterated by construction of a car park [5079 4192] at the Fox Hunter's Inn, proved a dextral displacement of 260 m. Fine-grained sandstones, generally massive or thickly bedded, are exposed in old quarries in and around Buttercombe Wood [504 416]. The single thickest sequence seen, 21 m, was at the north end of the wood [5069 4185], where dips are 10° to 25° southerly. Old quarries west of the railway show strong thickly bedded and massive, locally micaceous

Plate 6 River Barle near Simonsbath
Morte Slates dip at 60° southward in crags bordering the alluvium at Cornham Ford.
(A 13012)

fine-grained sandstones, dipping at up to 45°/180° to 200°, but in places steeper and sheared, with subordinate thinly bedded sandstones and siltstones and lenses of shale-pellet conglomerate [5071 4193], and 21 m of hard thick-bedded and massive fine-grained sandstone with a little thinner bedded sandstone [5069 4185]. A railway cutting [5025 4137] exposes thickly bedded and massive fine-grained sandstone, locally thinner bedded and sheared, generally dipping gently southwards; traces of current bedding show the strata to be right way up. An old quarry [503 412] east of the railway is cut in thickly bedded and massive fine-grained sandstones with some shale-pellet conglomerate.

Exposures behind Pembroke House [5178 4084] show rubbly brown and purple sandstone with ramifying veins of manganese oxide. Snow Ball Wood [520 410] contains much sandstone and an old manganese mine, Snowball Hill Mine (p. 74). Roadside exposures to the east, just south of Fullabrook [525 410], show purple and brown fine- to medium-grained micaceous and feldspathic sandstone with veinlets of manganese oxide, together with subordinate micaceous siltstone and some shaly strata. The disused Fullabrook Mine (p. 74) lies in a small east–west valley 1 km to the south-west. An old quarry [5310 3932] at Patsford shows 8 m of hard brown and grey thickly bedded and massive fine-grained sandstone dipping at 35° south-south-west. A broad rise north of Marwood [545 376] carries much rubble of purple sandstone and reflects Pickwell Down Sandstones in the core of an anticline.

Little Silver Quarry [549 403] (p. 60) is being worked in 105 m of bedded and massive, brown, grey and purple fine- to medium-grained sandstones with subordinate shale; the dip of about 45°/190° to 200° has locally been steepened by superficial creep, for instance high in the face at the northern end. Locally the sandstones are feldspathic, micaceous and with dendritic patterns of manganese staining. Current bedding shows the strata to be right way up. Some movement has taken place on bedding planes, with the generation of clayey silty fault breccias.

Bittadon to Bratton Fleming

The type locality of the 'Bittadon felsite', the tuff which marks the base of the Pickwell Down Sandstones, is a quarry [5502 4086] (Figure 2) alongside the driveway to Hewish Barton. The face shows 8 m of massive, slightly foliated, fine- to coarse-grained, rust-spotted rock characterised by broken feldspar laths. At its north side the tuff abuts against silty slates; at its south side thin quartz veins

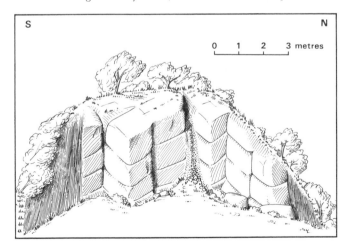

Figure 2   Quarry in tuff at Bittadon
This quarry is the type locality of the 'Bittadon felsite', the tuff which marks the base of the Pickwell Down Sandstones. About 8 m of massive tuff are exposed within slates and silty slates. Two small faults at the head of a small cleft trend north-west and dip steeply north-east

are associated with the junction, with brown, grey and purple silty slates and thin fine-grained sandstones, vertical and striking 110°. About 1 km to the east [560 406] the tuff forms a small feature with much surface debris.

Much purple fine-grained sandstone crops out in woods around Whitefield Barton [5575 3935]. Strata low in the formation are shown by small quarries [5728 4015; 5791 4003; 5842 4008] near Bowden Farm to comprise grey, brown and purple current-bedded fine-grained sandstones, locally micaceous and feldspathic, with a little siltstone and shale, dipping at 60° to 80°/170° to 175°. Higher beds around Viveham Mine [570 389] are exposed as fine-grained sandstones with quartz veins, hematitic fault breccias and manganese oxide. Roadside outcrops high in the formation at Muddiford show purple micaceous silty sandstones and silty shales overlain by buff sandstones, the beds dipping at 45° to 50° southward and with cleavage in the shaly parts inclined vertically or very steeply southwards [5628 3826]; and pale greyish green silty sandstone and thicker bedded micaceous fine-grained sandstone 2 m, overlain by purple and greyish green fine-grained sandstone with some siltstone 5 m, dip 55°/180° [5633 3825]. Exposures at a similar stratigraphic level near Plaistow Barton show sandstone beds up to 0.6 m thick with interbedded thin sandstones and siltstones [5754 3815].

The outcrop of the basal tuff band sweeps round the north-facing slopes of Churchill Down [594 403] and the tuff has been quarried close to the A39 road [5970 4008; 5986 4005]. Farther south, beds near the top of the formation comprise greenish grey thinly bedded fine-grained sandstones [5814 3776] north of Youlston Park, and similar strata with interbedded shales [5912 3790] south of The Warren; about 30 m of thickly bedded and massive grey and purple fine-grained sandstones dip at 65°/185° in an old quarry [5991 3784].

A small quarry [6102 3813] just south of Loxhore Cott shows 10 m of fine-grained and silty sandstones dipping at 80°/190°; shaly partings exhibit vertical cleavage planes aligned parallel to the strike. Roadside sections to the south display successively younger beds, dipping at 45° to 80° southward, as follows: thinly bedded and medium-bedded micaceous sandstones with shaly partings [6095 3780], overlain by thick-bedded and massive purple fine-grained sandstones with joints dipping steeply to between 280° and 310° [6094 3777], sporadic thickly bedded and massive purple sandstones [6093 3772 to 6088 3753], and thinner bedded fine-grained sandstones [6086 3745]. An east–west lane [6147 3764 to 6156 3764] at Lower Loxhore is floored by medium and thickly bedded sandstones with some shaly partings; the beds are vertical or dip steeply to north or south.

The basal tuff has been quarried [6324 3894] on the edge of woodland north-east of Smythapark, where a 4-m-wide face shows the rock to be massive with many broken feldspar laths. Overgrown outcrops at the southern end of the old platform at Bratton Fleming station show tuff [6413 3836] overlying Morte Slates and overlain by purple and grey sandstone, siltstone and shale [6412 3833], succeeded southwards along the old railway line by purple sandstone with subordinate siltstone and shale [6411 3829 to 6410 3824], sporadic sandstone [6409 3820 to 6408 3807], and south-dipping thickly bedded and massive fine-grained sandstone [6409 3804 and 6803 3794].

Haxton Down Lane contains exposures of purple and brown thinly bedded fine-grained sandstones with some shale in the top part of the formation [6457 3688 to 6468 3690]. The junction with the Upcott Slates [6427 3677] is seen in the lane from West Haxton to Lower Haxton; purple sandstones are succeeded by buff slates with a few sandstone bands.   EAE

Bratton Fleming to Darlick Moors

The tuff at the base of the formation can be traced as far east as a point [6583 3817] 350 m SW of Four Cross Way. East of here, fragments of tuff are met with at the same horizon, particularly between the last locality and the River Bray, and east of the Bray for just over 1 km to near Gratton [694 373], but in no case are they sufficiently abundant for the tuff to be mapped as a continuous bed.

There are several old quarries along the outcrop of the Pickwell Down Sandstones between Bratton Fleming and the B3226 Combe Martin to South Molton road. In one [6522 3757], purple fine-grained sandstone 2 m thick dips at 50°/185°; in another [6575 3714], a 0.08-m-wide quartz vein with limonitic cavities transects 3.5 m of purplish brown fine-grained sandstone with shaly partings. In an old roadside quarry [6634 3780] 3.5 m of purple thinly bedded sandstone dip at 70°/185°.

East of the Combe Martin road, 8.5 m of purple fine-grained sandstone dip at 85°/180° in Berryhill Quarry [6695 3717], and sandstone fragments abound in the soil between here and the River Bray valley. East of the Bray, around Lydcott, exposures of purple fine-grained sandstone are commonly found in the banks of the deeply-sunken lanes, and 2 m of greenish purple shale are exposed in the roadside [6967 3598].

An old quarry [7393 3609] at Span Head reveals 3.5 m of purple fine-grained sandstone with pale green shaly bands, and 800 m to the south-south-west, in Lyddicombe Bottom, the stream bed contains 10 m of green and purple thinly bedded fine-grained sandstone. The high ground of Long Holcombe is poorly exposed although resistant quartz-veined sandstone blocks abound and scattered old pits [753 357 to 759 357], marking the former extraction of stone used in the local dry-stone walls, are found to the west of Long Holcombe. Thickly bedded green and purple fine-grained sandstones can be seen in the stream bed north of Shortcombe [7661 3432 to 7674 3450], but east and south of here, towards Darlick Moors and North Molton Ridge, the outcrop of the formation is concealed by thick head and a thin mantle of peat.

East Buckland to Headgate

On the east side of the River Bray valley, green slaty sandstone crops out in the road [7011 3203] leading south from Walscott Farm, and the fields hereabouts bear a dense brash of grey, green, brown and purple fine-grained sandstone. A quarry at Stowford Brake [7117 3202] contains 6 m of medium and thinly bedded purple and red sandstone resting on 4 m of purple slate. At 230 m to the south, 3.5 m of medium and thinly bedded blocky purple micaceous fine-grained sandstone are exposed in another old quarry. About 200 m SE of this last quarry, 3.5 m of thick and medium bedded brown and greenish grey fine-grained sandstone dip at 30°/350°. On the east side of this valley, 200 m to the north-east along the strike from the last exposure, 4.5 m of purple very fine-grained sandstone dip at 15° to 30° [7144 3174]. Some 10 m of uniformly thickly bedded purple sandstone dip at 25° northward in a quarry [7162 3158] at Lower Barton; near Flitton Barton 1 m of mauve thinly bedded sandstone with shale partings rest on 1.5 m of green medium bedded sandstone, on 0.3 m of purple slaty sandstone [7165 3101]. The quarries at Barton Pits [723 321] are largely overgrown but debris of red and purple fine-grained sandstone and silty mudstone is abundant. At 500 m to the south, a quarry [7228 3153] shows 2.5 m of thickly bedded, hard, massive, fine-grained sandstone.

There are many exposures in the valley of the River Mole in the vicinity of Heasley Mill. About 4.2 m of thinly bedded mainly grey fine-grained sandstone dip at 50°/345° behind the Methodist Chapel [7380 3229], and in Crowbarn Wood 6 m of purple and green thickly bedded fine-grained sandstone with thin quartz veins can be seen [7383 3182]. An old quarry [7435 3136] in South Wood reveals a 10.5-m section in grey, green and purple, thickly bedded, fine-grained sandstone with pale greenish grey sandy shale partings, dipping at 45°/355°. On the higher ground to the east of Barcombe Down the head cover conceals the outcrop of the formation, but 2.5 m of weathered purple and brown rubbly sandstone crop out at Holywell Bridge [7674 3168]. Purple massive fine-grained

sandstone 4 m thick is exposed in a quarry [7621 3026] near Praunsley, and 2.5 m of purple and green thickly bedded fine-grained sandstone occur in a quarry [7823 3057] near Headgate.   BJW

Upcott Slates

Winsham – Marwood – Haxton

Silty slates crop out in numerous roadside exposures between Middle Winsham [499 389] and Lower Winsham [499 387]. Cleavage planes are vertical, trending slightly south of east, or dip 40° to 60° to north-north-east or south-south-west. Pale greyish green slates and silty slates typical of the formation are common in roadsides at and north of Beara Charter Barton [5214 3867 to 5245 3834], where cleavage planes are vertical or dip steeply south-south-west, and near St Michael's Church, Marwood [5429 3762 to 5446 3756].

Slates and silty slates sporadically exposed in the roadside [5677 3776 to 5696 3763] at Plaistow Mill show cleavage planes vertical or steeply inclined just east of south. The rocks are greyish green with some purple staining, and contain a few current-bedded silty bands and also silty nodules. On the west side of the Bradiford Water silty slates crop out immediately on the north side of a small quarry in Baggy Sandstones [5669 3749]. Similar rocks in the roadsides at Shirwell [598 374] show cleavage vertical or dipping steeply north or south.

East of Shirwell the belt of Upcott Slates commonly occupies a depression between the neighbouring sandy formations. Greyish green slates and silty slates crop out in an old pit [6061 3708] west of the River Yeo and in the side of a road [6096 3715 to 6109 3699] following the eastern edge of the alluvium. Similar rocks near Chumhill show cleavage vertical or inclined steeply to slightly east of south [6240 3695 to 6272 3670].

Extensive pavements of bare slates in and around Lower Haxton [6410 3666] show cleavage planes vertical or dipping about 70° north-north-west or 45° south-south-east. The lane thence to West Haxton [643 368] cuts through rocks low in the formation; slates and slaty shales just above the junction [6427 3677] with underlying Pickwell Down Sandstones contain scattered thin beds of red fine-grained sandstone.   EAE

Mockham Down Gate to North Radworthy

Between Benton and the River Bray valley, the Upcott Slates are poorly exposed, but, as elsewhere, they form a pronounced depression, commonly in arable land (the "Barley Vein" of local farmers). In this region the slates are rather shaly; rubbly pale grey shale is exposed [6592 3640] in the lane leading to Benton, and 0.5 m of pale brownish grey shaly slate in Mockham Wood [6818 3586]. East of the Bray valley, purplish grey sandy shale and slate debris is common near the base of the formation and the beds dip steeply west of south in an area bounded to the east by the north–south fault following the steep-sided valley between Holewater [7024 3520] and Newton Bridge [6935 3230]. Farther south the main body of the outcrop is characterised by yellow-weathering slate and shale fragments and pale grey siltstone rubble. About 1 m of yellow-weathered shale dips at 55°/180° at Moorman's Down [7081 3424] and 2 m of grey silty slate dip at 70°/180° at Yarde Gate Farm [7173 3431]. Field fragments and brash around Fyldon and North Radworthy are chiefly of yellow-weathering silty slates and shales, with a few greyish white siltstones.

East Buckland to North Heasley

The outcrop of the Upcott Slates in the Bray valley area east of East Buckland generally occupies the low ground and, although exposures are rare, the formation consists of grey and pale grey slates and shales, with scattered pale grey siltstones. East of the Bray valley, towards North Heasley, the siltstone constituent becomes more obvious. Some 3 m of laminated buff to olive silty shale dip at 70°/345° [7170 3300]. A network of land drains [723 327] 400 m S of Stitchpool revealed pale grey silty shale and shaly siltstone beneath a 2-m cover of head. Similar silty shales and greyish white siltstones crop out in many small laneside exposures in and around North Heasley.

Rabscott to Lambscombe

Between Rabscott Hill and Northland Cross the outcrop of the formation is marked by debris of brown and greenish grey silty and sandy shale, and immediately south of Portgate Cross is a road cutting [7122 2987 to 7122 2976] through greenish grey shale with sandstone bands up to 0.05 m thick; brown and greenish grey silty slaty shales crop out in the roadside [7152 2996 to 7169 2998] leading to Stonybridge Cross. Greyish green silty shale crops out in a small roadside exposure [7188 2980] at Stonybridge Hill, and around Oakford Cross [731 296] much pale grey shaly siltstone has been ploughed. To the east exposures are rare but the formation can be seen to consist of grey silty shales and slates with subordinate shaly siltstones.
BJW

Baggy Sandstones

Boode to Sloley Barton

A quarry [5055 3778] near Boode, now in use as a rifle range, is cut in 12 m of fine- to medium-grained thickly bedded and massive, grey and buff sandstones, locally micaceous, lying near horizontal or with a very slight dip to the west. These beds are near the top of the Baggy Sandstones. Small fractures commonly contain thin quartz veins, traces of flattened ferruginous nodules occur on joint planes, and some sandstone tops carry ripple marks. Rogers (1926) obtained remains of the fish *Holoptychius* from this locality. Dr D. E. Butler has recorded '*Cucullaea*' *unilateralis*, *Eoschizodus? deltoideus* and *Palaeoneilo* from the north-east face. He notes that '*C.*' *unilateralis* is the "*Dolabra*" which is abundant in Goldring's (1971) Rough facies and occurs also in his *D. yoyo* and *A. curvatus* facies.

Quarries in Lee Wood [5348 3741; 5364 3744; 5424 3745] show mainly thin sandstones with interbedded silty shales. Some ferruginous staining is evident, and buff medium-grained thickly bedded and flaggy micaceous feldspathic sandstones are present locally. An old pit [5420 3755] on the edge of the wood shows 5 m of purplish brown, thickly bedded fine-grained sandstones with ferruginous partings; another [5428 3757], just east of the wood, shows 9 m of grey and brown fine-grained sandstones with some siltstone and shaly beds and traces of shells. To the east, on the outskirts of Guineaford, a quarry [5500 3737] near the base of the formation exposes fine- to medium-grained, brown-buff banded, thickly bedded and flaggy micaceous feldspathic sandstones with traces of subordinate shales. Friable sandstones and siltstones crop out in the lane at Kennacott [5615 3745]. They lie at the base of the formation, as do the fine-grained feldspathic micaceous sugary sandstones with interbedded silty shales in a quarry [5669 3747] to the east.

The large Plaistow Quarry [568 373], due west of Sloley Barton and sometimes called Sloley Quarry, exposes about 100 m of strata probably just above the middle of the formation. The main south bay (Figure 3) is cut in interbedded thin fine-grained sandstones, siltstones and shales, with some thickly bedded and massive sandstones up to 8 m thick. Micaceous strata are common, and ripple marks show the (south-dipping) succession to be right way up. The topmost beds present, at the southern end of the quarry, comprise 20 m of shales and sandstones with thicker friable sandstones too soft to merit working. Strata between the south bay and the (disused) north bay comprise shales and sandstones faulted against the southern exposures. Strata seen in the quarry face are less well exposed alongside the road at the quarry entrance.
EAE

Faunas collected from various places in Plaistow Quarry are detailed in Table 5, columns C–G. Three are worthy of further mention. Towards the northern end of the quarry [5684 3731], loose blocks, apparently from the immediately adjacent face, yielded '*Cucullaea*' and bellerophontacean gastropods. The presence of '*Cucullaea*' suggests that the deposits are of near-shore origin and probably of Goldring's (1971) Rough facies. A varied assemblage collected from the northern end of the south-east face [5688 3728] is comparable with that of Goldring's *Lingula* facies, and suggests a fairly near-shore depositional environment. Bivalves of the genera *Palaeoneilo*, *Prothyris*, *Ptychopteria* and *Sanguinolites*, along with *Lingula*,

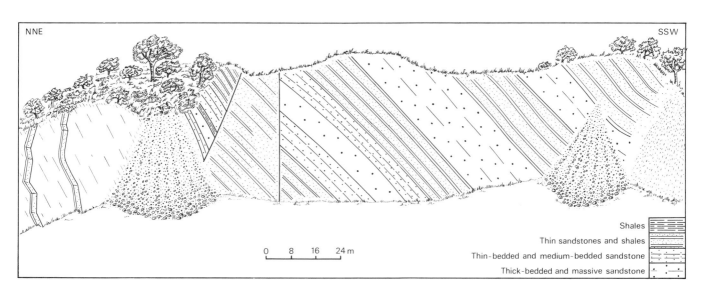

	Shales
	Thin sandstones and shales
	Thin-bedded and medium-bedded sandstone
	Thick-bedded and massive sandstone

Figure 3 Plaistow Quarry, Sloley Barton
South-dipping sandstones and shales of the Baggy Sandstones in the south bay of the quarry are cut by two minor strike faults

dominate this fauna, accounting for over 80 per cent of the identifiable remains collected. *Lingula* is an infaunal suspension feeder as, probably, was *Prothyris*; *Palaeoneilo* was an infaunal deposit feeder; *Sanguinolites* and *Ptychopteria* were probably semi-faunal suspension feeders. The fauna is reminiscent of that of earlier Upper Devonian strata encountered in the Willesden Borehole, in south-east England, where *Lingula*, *Palaeoneilo*, *Prothyris*, *Sanguinolites* and *Leptodesma* (a relative of *Ptychopteria*) were among the more common faunal elements in deposits interpreted (Butler, 1981) as of proximal inner shelf origin; as in the present instance, these were associated with bellerophontaceans and phyllocarid crustaceans. Rhynchonellaceans were prominent in the Willesden deposits, though there is some evidence that they disappeared with the slight approach of the shoreline or shallowing of the water; they were not found in these particular beds in Plaistow Quarry but do occur in strata exposed in the southern end of the south-east face [5684 3721], where they are associated with a sparse bivalve fauna and other remains including crinoid debris; *Lingula* was not recorded here. There is thus some slight evidence that the influence of the land was less on these deposits than on those at the northern end of the face. The assemblage does not, however, approach in character, that of the open-shelf, brachiopod-rich faunas found in later, Pilton A, strata, e.g. on the south side of Croyde Bay, on Down End and around West Pilton. DEB

Shirwell Cross to Stoke Rivers

Old pits [5899 3706; 5991 3702] and roadside exposures near Shirwell Cross show traces of the sugary micaceous sandstones characteristic of the Baggy Sandstones. Fine- to medium-grained medium and thickly bedded sandstones with interbedded shaly sandstones dip at 30° southward [6047 3701] at the head of a small valley on the eastern edge of Youlston Park. Thickly bedded and medium-grained sugary sandstones crop out in an old pit [6071 3669] to the south-east, and similar strata at the top of the formation are exposed [6097 3661] alongside the River Yeo. Hard micaceous rust-spotted sandstones have been quarried [6115 3675] beside the road immediately east of the river.

An old quarry [6161 3661] next to the old Lynton and Barnstaple Railway was cut in thickly bedded, fine- to medium-grained, micaceous sandstones with partings of thinly bedded micaceous sandstone in the upper part of the formation. Farther east and north-east along the track, silty and sandy shales with some thickly bedded fine- to medium-grained sandstones [6172 3665 to 6185 3666] lie at about the middle of the formation, and thinly and thickly bedded fine-grained sandstones, locally shaly, towards the base [6191 3672].

A lane south of Stoke Rivers cuts through thinly bedded rust-spotted fine- to medium-grained micaceous sandstones [6350 3496]. Small exposures at and east of Birch [6428 3594; 6445 3601; 6489 3612; 6499 3611] show fine-grained micaceous sandstones and silty shales. EAE

Mockham Down to Brayford

Brown sugary sandstone blocks and debris abound over the ridge forming Mockham Down. In a quarry within the earthwork at Mockham Down [6665 3585] 6 m of flaggy brown sandstone rest on 10 m of massive thickly bedded brown sandstone with ripple-marks. Brown sandstone debris is abundant around Brayford.

Elwell to North Molton

An old quarry [6616 3238] 500 m N of Elwell contains poorly exposed thinly bedded brown fine-grained sandstones with brown shaly partings, and the fields hereabouts bear a dense brash of brown sugary sandstone. However, the formation is poorly exposed between here and Great Rabscott. East of the River Bray, 3 m of

brown feldspathic fine-grained sandstone are exposed in a quarry [7047 2953] in Nadrid Copse, and the slopes south-west of North Lee [713 293] are covered with a dense brash of brown and buff flaggy sandstone and slaty shale. A quarry [7276 2918] 1 km SW of North Molton shows 12.3 m of thinly bedded to medium bedded brown and grey fine-grained sandstone with ripple-marks. A palynological report by Dr B. Owens on a sample from this quarry records *Punctatisporites spp.*, *Retusotriletes sp.*, *Calamospora spp.*, *Cyclogranisporites sp.*, *Hymenozonotriletes explanatus* and *Spelaeotriletes lepidophytus*. Dr Owens noted that most of the miospores recovered were difficult to identify even to generic level because of their opaque nature, that the previous known range of *H. explanatus* in the well-documented Belgian successions was from upper Tn 1a to lower Tn 3 and of *S. lepidophytus* from Fa 2d to upper Tn 1b, and that the joint occurrence of these two species indicated an uppermost Devonian age of upper Tn 1a or lower Tn 1b. BJW

Pilton Shales

Heanton Punchardon to Ashford

In the extreme west of the district, Goldring (1955) collected *Brachymetopus woodwardii* and *Piltonia salteri* from roadside shales and silty shales on Heanton Punchardon Hill [4980 3572] and assigned them to his Pilton C, regarding *P. salteri* as characteristic of this division. Shales nearby [4982 3572] have yielded a varied fauna (Table 6, column R). Much bare shale, with little sandstone, is exposed in Heanton Punchardon. *P. salteri* occurs in shales in an old cart track [5050 3582] north of the hamlet, and also in trackside exposures [5168 3583] in Ladywell Wood, where a small quarry shows mainly shales and silty shales with brown earthy decalcified fossiliferous lenses. Sandstones are not abundant in this western area, and the few thin mappable ones present are mainly confined to the northern, lower, part of the formation. Several small pits mark the digging of beds of fine-grained sandstone, commonly calcareous and locally micaceous [5104 3652; 5250 3642; 5333 3648].

The first of these, west of Luscott Barton, yielded *Whidbornella pauli*, and this species, along with *Hamlingella goergesi*, was also found [5278 3718] alongside a cottage at Pippacott (Goldring, 1957).

The ridge on which Heanton Punchardon stands runs east to Ashford and shows very little sandstone. Abundant exposure in and around the latter village shows shales and silty shales with some siltstones, dipping steeply to north and south. Goldring (1955) placed these beds in his Pilton B, and identified the following trilobites from roadside exposures: *Archegonus (Angustibole?) porteri* [5237 3543]; *A. (Phillibole) hercules* [5309 3556]; *A. (P.) duodecimae* and *A. (A.?) porteri* [5337 3533].

Shales and silty shales with a few thin calcareous sandstones, ferruginous silty nodules and densely fossiliferous bands dip steeply to slightly west of south on the estuary shore east of Chivenor [5102 3481 to 5100 3423]. More extensive outcrops along the shore to the east [5150 3490 to 5252 3480] comprise shales and siltstones with beds, lenses and nodules of fossiliferous grey and black limestone generally less than 0.15 m thick. These sections have been described in some detail by Goldring (1970), who recorded faunas indicative of both Pilton A and Pilton B. A small, probably Pilton A, fauna was collected during the survey from this area [5108 3461] and is detailed in Table 6, column M.

Allen's Rock to Sticklepath

Pilton Shales are brought to the surface in the core of an anticline which trends slightly south of east through Fremington Camp. They are sporadically exposed on the shore of the estuary in the neighbourhood of Allen's Rock [5029 3336], and also alongside Fremington Pill [5150 3288], and comprise shales and siltstones, locally pyritous, with a few limy bands.

The estuary shore north of Fremington Station carries outcrops of shales and silty shales disposed in close folds, with thin bands and lenses of fossiliferous limestone, some fresh but most brown and weathered [5172 3395 to 5173 3350]. This section was dealt with in some detail by Goldring (1970), who noted considerable faulting. He recorded a Pilton A3 fauna including *Whidbornella caperata* and *Phacops accipitrinus* towards its northern end, followed to the south by strata with Pilton B forms including *Archegonus (Phillibole) duodecimae* and *A. (P.) hercules*. To the south again, in the area around a basic dyke [517 336] he noted a Pilton C fauna including *Brachymetopus woodwardii* and *Piltonia salteri*, and at the southern end of the beach another occurrence of the Pilton B faunal division, with fossils including *Gattendorfia sp., Archegonus (A.?) porteri* and *Brachymetopus woodwardii*. The Devonian–Carboniferous junction thus lies within the northern part of this section; Goldring remarked that the change in fauna there is not accompanied by a change in lithology. Fossils collected during the survey from these exposures and their immediate neighbourhood are listed in Table 6, columns N, S, V and W.

Farther east, on the south side of the estuary, the alluvial flat terminates against a cliff up to 4 m high [5405 3291 to 5460 3273] bordering Anchor Wood, which exposes shales, siltstones and mudstones with scattered bands and lenses of limy siltstone or silty limestone; the dip is mainly steeply southward. Near the western end [5395 3296] shales yield a fauna which includes brachiopods suggestive of an early Carboniferous age (Table 6, column U).

Blakewell to Goodleigh

North of Barnstaple, small exposures of locally fossiferous shales and silty shales with thin sandstones are common, and several concentrations of sandstone beds can be mapped across the country. One of the two thickest trends east from just east of Varley [5499 3640] through Hartpiece Wood [5710 3605] to Sepscott [5915 3611]; the other from just east of Tutshill [5510 3540] almost to Brightlycott [5802 3538]. Small pits near Blatchford House show 2.5 m of grey fine-grained sandstone in grey shales [5657 3602] and beds of fine-grained calcareous sandstone up to 2 m thick in grey shales [5660 3592]; another [5744 3611], near Hartpiece, is in fossiliferous shales with a 0.6-m bed of calcareous sandstone; in a narrow gully [5867 3623] farther east vertical sandy beds have been dug out along strike. Goldring (1957; 1970) reported *Hamlingella piltonensis, Whidbornella caperata* and *Phacops accipitrinus*, indicative of Pilton A3, from exposures at Blatchford House [5668 3590]. In the more southerly sandstone, slightly larger pits near Playford Mill show faulted shales with hard micaceous, locally calcareous sandstone beds up to 0.4 m thick [5598 3508] (Figure 4), and faulted

and overfolded shales with calcareous sandstones and sandy limestones up to 0.4 m thick [5620 3512] (Top Orchard Quarry, Figure 5). A Pilton A1 fauna (Table 6, column B) was collected from a pit [560 351] during the survey. Top Orchard Quarry yielded a fauna including *Hamlingella goergesi* and *W. caperata* to Goldring (1957; 1970), who assigned the beds to Pilton A2, but they are here referred to Pilton A1 since *Whidbornella pauli* has been identified among material collected during the survey (Table 6, column C).

Roadside exposures immediately to the south, in the Roborough area, yielded the following trilobites (Goldring, 1955): *Brachymetopus woodwardii* and *Piltonia salteri* [5620 3483; 5695 3495], and *Archegonus (Angustibole?) porteri* [5695 3490]. An old quarry [5688 3532] shows fine-grained calcareous sandstone bands 0.4 m thick. Roadside exposures [5715 3503] east of Roborough show silty shales with earthy ferruginous fossiliferous bands and lenses, and an old quarry [5783 3485] is in shales and siltstones with thin limestones and fine-grained sandstones dipping at 50° to 65°/185° to 190° beneath hard massive calcareous sandstone. The quarry has yielded a small fauna (Table 6, column I) which is referred to Pilton A. A sandy development which passes south of Coxleigh Barton contains thin ripple-marked sandstones [5794 3507; 5866 3508].

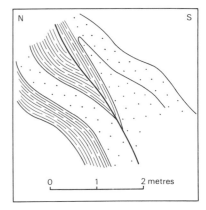

Figure 5 Quarry in Pilton Shales south-east of Playford Mill, Barnstaple
Shales and calcareous sandstones near the north end of the quarry show faulted overfolds. Sandstone (stippled); mudstone (ruled)

Bradiford–Barnstaple–Rumsam

Between Ashford and Barnstaple scattered sections, mostly at roadsides, show Pilton Shales dipping steeply to north and south. They have yielded *Piltonia salteri* (Pilton C) from alongside the Bradiford Water [5427 3408], and *Hamlingella piltonensis, Steinhagella steinhagei* and *Whidbornella caperata* (Pilton A3) from an old quarry [5502 3446] near the foot of Upcott Hill (Goldring, 1955, 1957, 1970). The quarry shows mainly shales and silty shales with weathered earthy limestone bands and lenses which contain the fossils. Immediately to the south, a small group of thin fine-grained sandstones crop out within micaceous silty shales. Pilton A3 faunas were collected from trackside and roadside exposures [5499 3446 and 5503 3448] during the survey (Table 6, columns G and H). Nearby laneside exposures show shales containing *Archegonus (Angustibole?) porteri?* and *Brachymetopus woodwardii* [5516 3469] and *Archegonus (Phillibole) duodecimae* [5535 3468] (Goldring, 1955). Laneside sections [5669 3392 to 5698 3399] east of Frankmarsh, on the north-eastern outskirts of Barnstaple, show shales and siltstones with scattered thin sandstones and thin beds and lenses of fossiliferous limestone. To the south a roadside exposure [5683 3308] of shales and silty shales with limy lenses on the eastern outskirts of Barnstaple has yielded *Hamlingella piltonensis* (Goldring, 1957).

The loop constructed to connect the railway from Taunton with Barnstaple Junction station passed through a deep cutting [564 318 to 569 321] in shales and silty shales with scattered siliceous bands and thin beds and lenses of limestone locally traversed by calcite veinlets. The siliceous bands accord with a position near the top of the formation, not far below the cherts and shales of the Codden

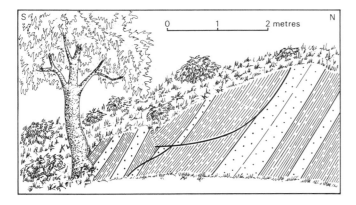

Figure 4 Quarry in Pilton Shales south-west of Playford Mill, Barnstaple
South-dipping shales and sandstones are cut by small strike faults

Hill Chert. Solitary corals, bivalves and crinoid debris were collected from the cutting [5692 3216] during the survey (Table 6, column T). Exposures of shales with earthy decalcified weathered bands [5734 3159] alongside Landkey Road west-south-west of Whiddon have yielded *Archegonus (Angustibole?) porteri* and *Brachymetopus woodwardii* (Goldring, 1955); these beds also lie very near the top of the Pilton Shales. North-east of Whiddon, silty and limy lenses up to 30 mm thick are exposed in a stream [5775 3185 to 5774 3198].

Riversmead to Charles

The road north-east of Riversmead has been cut into a hillside of shales with silty lenses and a little sandstone or limestone [599 357] which have yielded a Pilton A fauna (Table 6, column F). Exposures in Youlston Old Park show shales and silty shales with bands and lenses of siltstone and a little calcareous sandstone [6004 3634], and silty shales containing a 2-m bed of hard calcareous sandstone, dip 65°/350° [6079 3581]. Fossiliferous shales, silty shales and siltstones, locally nodular, with scattered thin sandstones, are well exposed in roadsides between Chelfham and Chelfham Barton [6118 3572 to 6171 3606].

Quarries near Snapper show a shaly succession with subordinate sandstone beds and limestone lenses [5948 3473], and thickly bedded sandstones with limestones and shales [5943 3448]. The latter locality has yielded a fauna including *Phacops (Omegops) accipitrinus accipitrinus* and productellid brachiopods indicative of Pilton A1 or A2 (Table 6, column E). A lane [597 341 to 599 343] on the western outskirts of Goodleigh cuts through shales and silty shales with thin fossiliferous limestones; a Pilton A fauna was collected [5986 3427] during the survey (Table 6, column J).

Between Goodleigh [599 341] and Hutcherton Down [639 330] the sandstone beds within the Pilton Shales tend to be fairly thin, commonly in packets up to 10 m thick which form well-defined features. Locally the sandstones have been quarried, but exposures of the shales tend to be limited to roadside banks, and many contain the characteristic rusty-weathered fossiliferous bands. At Bradninch Cross [612 332] a WSW–ESE-trending ridge is capped by a dense brash of fine-grained calcareous sandstone; probably the dip is southerly, as silty shale in a small exposure [6082 3335] dips at 65°/170°. Rusty-weathered silty shale with traces of brachiopods and crinoid ossicles crops out in the stream bed [6095 3320] 200 m to the south-east. About 2 m greyish brown shaly sandstone dip at 70°/345° in a small quarry [6151 3445] at the edge of Dean Wood.

A quarry [6361 3415] near Great Beccott shows thinly bedded brownish grey cross-bedded sandstone. Farther south, in an exposure [6369 3393] in Stonepark Copse, 6 m of brown silty and sandy shale dip at 85°/012° immediately beneath a sandstone bed. In another quarry [6453 3470], in Yarde Wood, 3 m of brown silty shale with rusty-weathering bands containing fossil traces dip at 70°/175°.

A prominent ridge trends east-north-east from near Yarde Cross, and the field fragments contain a high proportion of hard grey muddy siltstone and fine-grained sandstone. North of this ridge, brown silty shales crop out in small roadside exposures [665 352], and 3 m of brown silty shale with sandstone ribs up to 0.05 m thick dip at 30°/165° [6725 3474]. An old quarry [6588 3337] at Whitsford, south of the ridge, is largely overgrown and filled in, but traces of brown silty shales with thin (up to 0.05 m) calcareous siltstone bands can be seen. The proportion of sandstone and siltstone to shale within the formation increases gradually eastwards from here towards the River Bray valley, and extensive quarrying is carried on north of Charles (Plate 7), where over half the succession is locally composed of mainly thinly bedded sandstone. One large quarry [687 340] contains 30 m of hard grey fine-grained sandstone with hard grey silty shale in alternations up to 2 m thick. To the south another large quarry [687 336] reveals 6 m of thinly bedded

hard grey fine-grained sandstone on 15 m of alternating sandstone and shale similar to that exposed in the previous quarry, on 4 m of thickly bedded brownish grey fine-grained sandstone. The quarry [6900 3337] in Thorncleave Wood is in a 10-m sequence of thinly bedded alternating hard grey fine-grained sandstones and hard grey silty shales, and 240 m to the west-south-west Mogridge's Quarry [6879 3327] is cut into 3 m of brown-weathered cross-bedded fine-grained sandstone. EAE, BJW

Faunas collected from the working quarries near Charles are grouped together in Table 6, column A. These were collected from decalcified bands and closely associated sandstones representing only a small part of the sequences seen in the quarries. They have little in common with the typical Pilton Shales assemblages examined but are strongly reminiscent of the overall faunas obtained from the Baggy Sandstones and listed in Table 5. DEB

Barnstaple – Yarnacott – East Buckland

Aggregates of thin sandstone bands within the Pilton Shales pass south of Acland Barton and are exposed as brown fine-grained sandstones with some fossiliferous limestone [5971 3247], thinly bedded brown fine-grained feldspathic sandstone [5901 3246], and sandstone beds up to 0.25 m thick with interbedded shales and siltstones [5976 3226].

Old quarries [6006 3159; 6009 3159; 6021 3169] south-west of Harford show mainly silty shales. A spoil heap [6031 3167] in the valley immediately east of the easternmost quarry contains shale, some of it black and carbonaceous, with limestone and ferruginous sandstone. The methodical separation and disposal of waste suggests the extraction of limestone rather than small-scale pitting for sandstone; it seems likely that small lenses of limestone within shales have been completely worked out. Another small overgrown pit [6046 3154] and spoil heap [6051 3155] lie farther east. A small disused quarry [6128 3223] in Little Silver Wood, 1 km ENE of Harford, shows dark grey shales with thin calcareous and silty lenses; *Hamlingella goergesi* and *Whidbornella pauli* have been recorded (Goldring, 1957).

There are many small exposures of rubbly silty shale in the banks of the roads leading north and east from Yarnacott, amongst them 5 m of brown silty shale dipping at 75° southward [6256 3092], and others in the bed of the stream leading south-south-west from Taddiport. A roadside exposure [6251 3088] north-east of Yarnacott has yielded a Pilton A fauna (Table 6, column K).

The thick sandstones seen north of Charles do not persist across the anticline to the south into the area around West Buckland. In this vicinity a stream section [6524 3223] 0.5 km W of Stoodleigh reveals brownish grey shales with some rusty-weathering shelly bands dipping at 70°/170°, with a few calcareous siltstone and sandstone bands. At West Buckland Cross [6579 3161] grey thinly bedded papery shales dip at 10° to 60°/195°, and exposures of brownish grey shales near West Buckland School show dips of 85°/180° [6626 3154] and 80°/175° [6677 3140]. Towards the River Bray, hillsides south of East Buckland, south-east of Huxtable Farm, and east of Crossbury bear debris of greyish brown shale with calcareous siltstone and sandstone fragments.

Bydown House to Hacche Moor

Pilton Shales in the core of an anticline occupy a small east–west valley south of Bydown House [623 294], extending south of east, although cut by faults, through Kerscott [632 294] to Yollacombe Plantation [642 290]. A second similar outcrop and anticline trend north of east through Irishborough [634 282] and Tower Farm [640 284]. The two folds coalesce, and a single anticline in Pilton Shales trends south of east to pass immediately south of Castle Hill and thence beneath Stag's Head [676 278], Clatsworth [682 282] and Hacche Moor [716 273] to just east of the old South Molton Station. Exposures are few except in the River Bray and the River Mole.

Workable thicknesses of limestone are rare in the Pilton Shales, but limestone has been dug from two old quarries near the top of the formation at Leary Barton. One large quarry [6477 3000] contains shales with thin beds, lenses and rubble of grey sandy limestone and dark grey crystalline limestone. The limestones are locally weathered to brown rottenstone from which fossils stand out in relief. Similar shales with grey calcite-veined limestone occur in another quarry [6445 2972] to the south-west; remains of old limekilns are present at both sites. The two quarries are about 350 m apart, measured at right angles to the strike, and it is possible that they mark the same lenticular limestone in the opposite limbs of a fold. Good exposures of fossiliferous shales and silty shales with limy bands occur in an old railway cutting [6545 2945 to 6568 2946] near Leary Moors, and the remains of several small pits [665 291] and a limekiln in a wooded valley to the east attest to the presence of limestones.

Numerous exposures occur alongside the old railway track from Filleigh Station [674 293] to north of North Aller [696 282], all showing typical shales and silty shales with silty calcareous lenses and beds. Old dumps [6850 2855 to 6885 2848] alongside a stream south of Bremridge Wood containing mainly shale with quartz, galena, arsenopyrite and siderite. Grey shales and silty shales exposed near Snurridge dip at 85°/010° [7030 2831] and are vertical, trending 125° [7042 2814]. In Hacche Wood, an exposure [7206 2809] reveals 4 m of brown-weathered grey calcareous splintery fine-grained sandstone dipping at 42°/185°; the beds are quite hard and quartz-veined, and have a nodular appearance. They contain a Pilton A fauna – see Table 6, column L. Eastwards from here to the limit of the district the sandstones largely peter out and exposures show typical Pilton Shales, as for example in the stream section near Burwell [7509 2792 to 7508 2776], where greyish brown shales dip southerly at 45° to 65° and contain thin highly weathered rusty fossiliferous bands. In the lane [7514 2735] 500 m S of the stream section, grey silty shales yield a variety of fossils (Table 6, column D) indicative of Pilton A, probably A1. Just west of Reach Crossing, on the old railway, another old quarry [769 267] probably marks a limestone working. Exposures are of fossiliferous shales, but a good deal of limy rubble occurs. A smaller quarry [7745 2647] farther east may also have been opened for limestone. Both pits lie near a fault separating Pilton Shales from Upper Carboniferous rocks, and are near the top of the Pilton Shales.   EAE, BJW

Plate 7 Sandstones in Pilton Shales near Brayford
Sandstones with shaly partings dip at 45° northward in the northern limb of an anticline. (A 13016)

CHAPTER 3

Carboniferous

GENERAL ACCOUNT

LOWER CARBONIFEROUS

Codden Hill Chert

The Codden Hill Chert (Plate 8) overlies the Pilton Shales and its relatively narrow outcrop extends from the Taw estuary near Fremington, past Swimbridge to near East Marsh, north-east of South Molton. A second outcrop has been traced on the south side of an east–west anticline from South Molton westwards to Dinnaton, south of Swimbridge; thence it continues around a syncline and westwards in the core of the east–west anticline to pass south of Bishop's Tawton and end near Nottiston.

The succession is commonly predominantly shaly, but in many places it contains a sequence mainly of cherts. Generally these cherts are both underlain and overlain by Lower Carboniferous shales, and they form sharply defined steep-sided ridges of great assistance in delineating the formation in the field. Lenticular limestones occur at several localities; usually they lie in the upper shales but in some places where no mappable cherts occur they are present both low and high in the sequence. A generalised succession in which all the main subdivisions were present would be dark grey or black shales, locally siliceous and with a little limestone, overlain by creamy-buff-weathering cherts with a few shales, overlain by grey or black shales, locally siliceous and containing large lenses of dark grey limestone. The total thickness would be about 250 m, with the central cherts forming the thickest subdivision.

All these sediments originated in quiescent, probably comparatively shallow, waters associated with a shelf sea. A coastal plain lay over the present South Wales, and deeper water to the south. Slow accumulation of muds brought by sluggish rivers from a land of fairly low relief was followed by development of radiolarian cherts, perhaps in lagoons similar to those suggested for cherts near the base of the Millstone Grit in South Wales (Dixon and Vaughan, 1911). Further deposition of muds followed, and the formation of local lagoons within the shelf sea facilitated the development of lenticular limestones.

Early work on the Codden Hill Chert was comprehensively summarised by Hinde and Fox (1895), who described the microscopic characters of the cherts and also their abundant radiolaria and other fossils. They followed Phillips (1841) in placing the cherts above the grey shales and limestones. Ussher (1892), however, had been uncertain on this point and preferred to group the beds in a single 'series'. He later (1901) divided his Lower Culm into Basement Beds (shales), Codden Hill Beds (mudstones, cherty shales and cherts) and Limestone Series (shales and mudstones, locally cherty, with limestones), a succession now broadly confirmed in North Devon. Fossils collected by Hamling and Rogers led Hind (1904) to a similar conclusion about the position of the cherts.

Vaughan (1904), working on corals and brachiopods supplied by Hamling from quarries between Barnstaple and South Molton, concluded that fauna from the west, around Codden Hill and Landkey, pointed to the *Zaphrentis* Zone or upper *Cleistopora* Zone.

A report on a field meeting in north Devon (Hamling and Rogers, 1910) referred to a number of fossil localities. Garwood (*in* Evans and Stubblefield, 1929) thought that there was a break in succession between the top Pilton Shales and the Codden Hill Chert; he assigned the former to the *Zaphrentis* Zone and equated the latter with the topmost Viséan strata of the Bristol area. Richter and Richter (1939) quoted Stubblefield as placing the Codden Hill Chert not younger than the German goniatite zone IIIβ, the top of which closely equates with the top of the British zone P_1 (Hudson and Cotton, 1945).

Plate 8   Codden Hill Quarry
Steeply dipping banded cherts of the Lower Carboniferous. (A 11827)

A major contribution in recent years to the study of the Carboniferous rocks immediately south of the Taw estuary was that of Prentice (1960a), who sought to distinguish two facies of Lower Carboniferous rocks brought together by a number of thrusts. He regarded the successions at Fremington and Codden Hill as of not completely identical age, although the '*Goniatites spiralis*' Bed occurred at the top of each facies and was considered to be of basal P_{2a} (IIIβ–γ) age. The main difference is that, at Fremington, cherts and cherty shales make up no more than one-third of the formation, whereas at Codden Hill they form a thick succession beneath shales.

Specimens in the North Devon Athenaeum, Barnstaple, from the Fremington area were listed by Prentice (1960a) as *Girtyoceras* aff. *brueningianum* from Fremington and *Goniatites* aff. *falcatus* from Fremington and Bickington; *G. falcatus* is indicative of zone P_{1b} in the north of England. *Posidonia becheri* occurs both low in the succession and near the top. Prentice (1960a) assigned trilobites from two levels in his Codden Hill facies to horizons close to the junction of zones II and III (ie. close to the base of P_{1a}) and within zone IIIβ. The lower source also yielded goniatites suggesting zone IIγ. He concluded from the trilobites and goniatites recorded, and from the known range of *P. becheri* in northern England, that strata ranging from P_{1a} to P_{2a} were present in both facies, but that 9 m of strata below the lowest trilobite horizon of his Codden Hill facies probably represented beds of zone II.

Detailed mapping of the whole Lower Carboniferous outcrop, from Fremington to its disappearance against the Brushford Fault (p. 59) east of South Molton, shows that the rocks are disposed in large upright folds (Plate 9) broken by numerous faults but no major thrusts. Cherts are common in the central tract, between the rivers Taw and Bray, but subordinate in the west and east; possibly this indicates the limits of suitable lagoonal depositional environments. It is likely that any apparent difference in age-range between one local succession and another is negligible or reflects incomplete faunal records.   EAE, BJW

UPPER CARBONIFEROUS

Upper Carboniferous rocks occupy a triangular area in the south-west and south of the Barnstaple district. The interpretation of stratigraphical relationships and distinctions within the sequence is largely based on eastward projections of observations made on the coastal sections south of Westward Ho!

Plate 9 Topography of Carboniferous rocks near Swimbridge
Indiwell Farm (centre) is situated on Crackington Formation shales and sandstones in the core of a syncline between two ridges of Codden Hill Chert. (A 11836)

Shales with turbidite sandstones of the Crackington Formation overlie the Codden Hill Chert and extend upwards to about 200 m above the *Gastrioceras listeri* horizon. They include the Limekiln Beds and Instow Beds of Prentice (1960a) and the Westward Ho! Formation of De Raaf and others (1965). The base of the overlying Bude Formation is taken at the base of the lowest of the massive sandstones typical of the formation. However, lithologies within the Crackington and Bude formations are commonly indistinguishable in the field in areas of poor exposure, and it is probable that some repetition by folding has gone undetected.

The Bideford Formation is equivalent to the Bideford Group of De Raaf and others (1965), and the Northam and Abbotsham Beds of Prentice (1960b), it comprises a paralic cyclic succession containing massive cross-bedded sandstones, and is equivalent in age to part of the Bude Formation into which it passes laterally. No Bideford Formation strata are present east of about South Molton.

It appears that the shallow sea of Lower Carboniferous times was succeeded by a deeper basin whose submarine slopes were periodically swept by turbid sediment-laden currents, initiated in response to disturbances such as earth tremors or sudden influxes of floodwater. Subsequently, in Westphalian times, deltas advanced southwards into north Devon and fresh-water and brackish-water sediments were deposited; discrete bands of dark grey and black marine shales mark spasmodic deepening of the sea. Sands, silts and muds that accumulated on the delta tops in a subaerial or subaqueous environment are now represented by the Bideford Formation, characterised by massive channel-fill sandstone (Edmonds and others, 1979). Sediments laid down at the same time in deeper water now form the Bude Formation. Mixed facies were deposited on the upper delta slopes. Thick sandstones of the Bude and Bideford formations show a similar grain-size distribution (Freshney, Edmonds, Taylor and Williams, 1979), although the latter are slightly coarser, and it is possible that sands on the delta tops were spread seawards down the delta slopes by interlacing fans of river-water channels. This would account for the present passage from Bideford Formation to Bude Formation, as a reflection of continuous sedimentation and redistribution over the frontal areas of delta lobes.

Crackington Formation

The Crackington Formation comprises greyish green fine-grained sandstones, rarely more than 1 m thick, bearing a variety of sole-structures and locally showing grading and lamination, in a sequence of siltstones, mudstones and shales. The succession is more argillaceous than the Crackington Formation south and south-west of the district (Freshney, Beer and Wright, 1979; Freshney, Edmonds, Taylor and Williams, 1979) and somewhat akin to the Westphalian strata below the *G. listeri* horizon, the Wanson Beds of Mackintosh (1965). Evidence of flow direction is variable, but probably most currents came from the north.

Moore (1929) recorded *Reticuloceras* in strata near the mouth of Fremington Pill which are known to belong to the Crackington Formation. Prentice (1960a) located both of Moore's localities. He identified *R. reticulatum* from the foreshore [5135 3325] and *R. gracile, R. sp.* and *Dunbarella rhythmica* from the base of the small cliff [5135 3324].

Moore's (1929) recognition of zones R_1 and R_2 is confirmed as R_{1a} and R_{2a}. Prentice (1960a) also collected specimens of *Reticuloceras* from a stream section [5251 2968] east-north-east of Litchardon. He noted a further fossiliferous locality in an old quarry [5077 3115] in Bickleton Wood, where he recorded *Gastrioceras* cf. *carbonarium, Anthracoceras arcuatilobum, Metacoceras sp.* and *Dunbarella papyracea* var. A, and he equated this fauna with the *G. listeri* horizon near the top of the Crackington Formation. Goniatites collected from Venn Quarries [581 306] by Dr Clive Nicholas of EEC Quarries Ltd have been identified by Dr W. H. C. Ramsbottom as *Homoceras*, possibly *H. beyrichianum* (H_{1b}).

Bideford Formation

The Bideford Formation contains sequences of thin sandstones, grey muddy siltstones and pale to dark grey or black shales indistinguishable from similar rocks in the Crackington and Bude formations. With them, however, occur fine- to medium-grained massive or thickly bedded feldspathic cross-bedded sandstones which weather to a sugary texture. These sandstones give rise to well-marked east–west ridges which are a striking feature of the Bideford Formation topography. Silty and sandy mudstones also occur, but it is not possible in this inland area to distinguish the pattern of cycles recognised on the coast in the adjacent Bideford (292) district, although the southernmost sandstone in the south-western part of the Bideford Formation outcrop may be equivalent to the Cornborough Sandstone of the Bideford district. The feldspathic sandstones become reduced in thickness and number towards the east.

Bude Formation

The Bude Formation consists of thickly bedded and massive sandstones, with thinner sandstones, siltstones, mudstones and shales. Sequences of thin sandstones and argillaceous beds commonly resemble the Crackington Formation. The thicker sandstones are generally slightly coarser grained, become distinctly more friable on weathering and are the main criterion by which the formation is identified in the field. Some cross-bedding is evident locally, but there is no good evidence for the general direction of flow of depositional currents.

Sandstones striking east–west across the valley of the River Bray in the vicinity of Bradbury Barton include a number of massive and thickly bedded aspect, typical of the Bude Formation. A few are medium grained, feldspathic and weathered to a sugary texture, somewhat reminiscent of the Bideford Formation, and traces of such rocks occur, for example, south-east of Bradbury Barton [6701 2628] and near High Bray [6795 2654].

Immediately south of the southernmost Bideford Formation sandstone in the south-western extremity of the district, around Alverdiscott, several old shafts and workings occur, and are on the eastward extension of the belt of workings in the 'culm' coals traced across the adjacent Bideford (292) district. East of here, old shafts and adits occur around Somers [556 256] and in Hawkridge Wood [604 252]. Both the latter areas lie within the Bideford Formation outcrop, and indicate worked 'culm' beds either in the topmost part of the Bideford Formation or in unmappable synclinal areas of basal Bude Formation beds.   BJW, EAE

STRATIGRAPHICAL PALAEONTOLOGY

Data from new material obtained during the re-survey is outlined below and full locality and faunal information is available from the Palaeontology Research Group at Keyworth.

Important Dinantian discoveries include the goniatite *Protocanites sp.* in transitional strata between the Pilton Shales and Codden Hill Cherts at Castle Hill [6762 2868] indicating that the Pilton Shales terminate within the Tournaisian.

The Tournaisian/Viséan boundary probably lies within the lower part of the Codden Hill Cherts as late Chadian faunas occur in this unit at Templeton Quarry [5449 2974]; Codden Hill Quarry [5697 2970] and Park Gate Quarry [5570 2971], and include the goniatites *Ammonellipsites*, *Eonomismoceras* and *Merocanites*, accompanied by the trilobites *Cystispina (Spatulina) spatulata* (Woodward) and *Archegonus (Phillibole) coddonensis* (Woodward). Arundian–Asbian faunas are poorly represented possibly due to stratigraphical attenuation. Late Brigantian faunal horizons near the top of the Codden Hill Cherts include *Goniatites* cf. *koboldi* Ruprecht (P_{1d}) at Bestridge Quarry, Swimbridge [6251 2981]. A rich P_{2a} fauna was recovered from Clampitt, Fremington [5139 3298] with *Goniatites granosus* Portlock, *Neoglyphioceras caneyanum* (Girty) accompanied by the trilobites *Archegonus (A.) tevergensis* Gandl, cf. *Kulmiella sp.* and cf. *Pseudospatulina*. The youngest Brigantian fauna was obtained from the roadside exposure near St Johns Chapel, Tawstock [5315 2975] and includes *Sudeticeras* cf. *adeps* Moore and *S. ordinatum* Moore, indicating a P_{2b} horizon.

The Namurian of the Crackington Formation only yielded early Kinderscoutian faunas although a complete Namurian sequence is probably present. At Clampitt, Fremington [5131 3321] the horizons of *Reticuloceras subreticulatum* (Foord), *R. todmordenense* Bisat & Hudson and *R. aff. eoreticulatum* Bisat emend. Bisat & Hudson were recovered, indicating a late R_{1a} to basal R_{1b}? age.

The characteristic basal Westphalian goniatite *Gastrioceras subcrenatum* Schmidt occurs at an old quarry [5076 3118] near Collacot Farm, Fremington. Younger Westphalian horizons are not represented in the recent collections.   NJR

DETAILS

LOWER CARBONIFEROUS

Codden Hill Chert

Fremington to Barnstaple

At Fremington Station [5156 3329] the Codden Hill Chert comprises 70 m mainly of shales and cherty shales dipping at about 45°/200° in the northern limb of a syncline and yielding *P. becheri* from exposures on the foreshore. Shales, mudstones and cherts in the southern limb of the syncline crop out on the shore of the estuary around 300 m WSW of the mouth of Fremington Pill and similar rocks in the banks of the Pill immediately south of Pill Cottage [5134 3306] show fractures, overfolds and contortions. They form a narrower outcrop than do the corresponding strata at Fremington Station. *P. becheri*, *P. membranacea* and repetition of the '*G. spiralis*' Bed were noted by Prentice (1960a) in the western bank of the Pill.

These Codden Hill Chert strata also form the northern limb of an anticline. They are repeated in the southern limb, which runs beneath Fremington Camp; exposure is poor but traces of an old pit [5050 3284] occur south of Saltmill Duck Pond and the beds have been noted in the foreshore of the estuary to the west (Edmonds and others, 1979). An old quarry [519 332] near Penhill was opened to work one of the limestone lenses which are common within the Codden Hill Chert. The face shows about 4 m of the thickly bedded calcite-veined grey limestone with subordinate shale, siltstone and chert.

Between Fremington Camp and Barnstaple the Codden Hill Chert crops out in two belts separated by Crackington Formation and locally displaced by north-westerly and north-easterly faults. The northern belt maintains south-south-westerly dips and a probable thickness of about 100 m. The southern belt crops out south of the A39 road at Fremington, along the ESE-trending valley at Muddlebridge, and beneath the A39 road through Bickington. At Fremington it shows faults and some inversion. Pits [5191 3251; 5207 3247; 5226 3246] near Muddlebridge contain shales, cherts and dark grey limestone, and another old quarry [5222 3240] immediately to the south was opened in bedded cherts. This southern belt marks an anticlinal crest affected by minor faults and small-scale folds.

Urban development on the western edge of Barnstaple now surrounds a flooded limestone quarry [5427 3227] in which thin-bedded limestones and cherts are exposed; at the eastern end of the pit the strata are disposed in open folds with east–west axes. A dextral wrench fault runs south-westwards through the western outskirts of Barnstaple towards Tawstock, and Codden Hill Chert to the east trends south of east beneath Lake but is largely hidden by boulder clay. A large quarry [554 315] at Lake marks the working of another lenticular limestone. Only siliceous shales and siltstones with a few thin limestones now crop out, but the spoil heap shows, in addition, more massive grey limestone, sandstone and chert. If Prentice (1960a) was correct in thinking that shales and siltstones exposed in the bottom were Pilton Shales, their occurrence is probably due to the minor folds which are almost certainly present; however there is no faunal evidence and lithological distinctions are not clear.

Rumsam to Swimbridge

East of the River Taw mappable cherts occur between Rumsam and Landkey. The northern belt of Codden Hill Chert thickens eastward from about 100 m to perhaps 250 m. It comprises shales overlain by cherts overlain by shales, with thicknesses of 15 m, 70 m, 15 m respectively in the west, and possibly 30 m, 150 m, 70 m near Venn Quarries, where the upper shales contained limestone lenses up to 20 m thick, now largely quarried away, which have yielded *Posidonia*. This thicker succession continues to Swimbridge, where limestones in the upper shales have been worked.

A nursery garden beside Venn Road occupies the site of an old quarry [5724 3134] in 6 m of grey and brown banded cherts with interbedded shales. Similar strata have been quarried [5789 3125] on the crest of the ridge farther east. Large limestone quarries [5808 3091 to 5879 3078] to the south-east have been largely filled with waste from the active workings of Venn Quarries; a few exposures of shales and silty shales remain, and thick-bedded limestones with shales, siltstones and cherts crop out in a syncline at the western end. A small quarry [6060 3040] south of the A361 road near Newland House contains cherts in beds commonly around 0.15 m thick, with a few thicker beds of hard grey fine-grained limestone. Similar rocks occur on strike in another pit [6079 3029]; they have been extensively exploited for limestone in Marsh Quarries [613 302], but exposures have been largely hidden by tip. Cherts quarried

[6155 3028] (Figure 6) north of the A361 road occur in thin beds with some shales and limy bands in a disturbed 8-m succession. About 100 m ESE some 3 m of banded cherts with shale partings crop out in an old pit [6166 3024]. A larger quarry [6201 3013] (Figure 7), on the edge of Swimbridge, shows 20 m of thin-bedded cherts with shaly partings and limy bands, much folded and faulted. Bestridge Quarry [625 299] exposes shales and silty shales with fine-grained earthy limestone and grey crystalline limestone, yielding *Posidonia*.   EAE

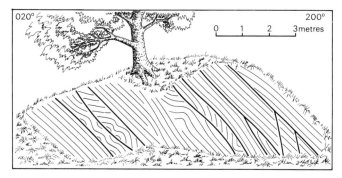

Figure 6   Quarry in Codden Hill Chert west of Swimbridge
Thinly bedded cherts with some shales and calcareous bands are folded and cut by several minor strike faults which commonly follow bedding

Nottiston to Hearson Hill

Strata at Lake (p. 43) dip south beneath Crackington Formation but reappear at Tawstock in the crest of an anticline. This southerly occurrence contains the westernmost development of separately mappable cherts, which lie below Lower Carboniferous shales.

Dark grey silty shales crop out in a stream bed [5251 2968] in Nottiston Copse; Prentice (1960a, p. 278) recorded *Reticuloceras sp.* from here in the shales overlying the cherts (his Limekiln Beds). Farther east, a track section south-east of Nottiston [5316 2974] reveals the top of the cherts overlain by black siliceous shale from which Prentice recorded *Posidonia becheri*, overlain by black shale with goniatites (his 'Goniatites spiralis' horizon), from which he identified *Neoglyphioceras spirale*, cf. *N. spirale*, and cf. *Mesoglyphioceras granosum*. East of Eastacombe, Templeton Quarry [5432 2972] reveals grey banded chert with thin rather shaly bands, containing scattered crinoid fragments, and Prentice (1960a) recorded *Phillibole (Phillibole) coddonensis* from his higher fossiliferous horizon. In Park Gate Quarry [5552 2970] grey and black cherts with shaly bands are exposed; Prentice (1960a) obtained *Cyrtosymbole (Waribole)* cf. *aequalis* and *Carbonocoryphe* aff. *bindemanni* from his lower fossiliferous horizon and *Phillibole (Liobole) glabra glabra* and *P.?* (*Cystispina*) *spatulata* from scree below. Near Hillside about 0.5 m of banded chert with a dip of 45°/170° [5449 2526] lies south of the main outcrop. This exposure caused Prentice (1960a, p. 280) to postulate his Tadiport thrust, but it is here considered to be part of the southern limb of a strike-faulted anticline.

The main quarry [5696 2972] at Codden Hill (Plate 8) shows cherts in the northern limb of an anticline in Lower Carboniferous rocks which trends east from Tawstock; about 20 m of mainly hard, pale and dark grey colour-banded chert, with pale grey to white siliceous shale bands up to 0.05 m thick, dip at 65°/355°. In Overton Quarry [5735 2944] 10 m of rather rubbly weathered pale grey cherts are exposed, and in another quarry [5829 2942] south of Codden Beacon, 12.2 m of grey and black banded chert dip at 80° southward. East of Codden Hill the chert outcrop divides to form two ridges as lower beds are exposed in the core of the anticline, one forming Hangman's Hill and the other Hearson Hill; in the southern ridge [5964 2938] pale and dark grey banded chert dips at 75° southward. Thinly bedded cherts have been quarried near Bableigh and near Hannaford [6030 2984] ; the latter locality is probably the place where Hinde and Fox (1895) recorded trilobites

Figure 7   Quarry in Codden Hill Chert at Swimbridge
A large quarry on the northern edge of the village shows thinly bedded cherts with shaly partings and calcareous bands, much folded and faulted at the southern end of the face

and brachiopods. The rubble at Hannaford Mine [607 296] suggests thin shales of the Codden Hill Chert overlying Pilton Shales. Old pits on the south side of Hearson Hill show shales and siltstones with fossiliferous limy lenses (Pilton Shales) 1.2 m, overlain by cherts and siliceous shales 2 m [6037 2932], and shales, siltstones and cherts [6092 2928].   BJW, EAE

Swimbridge to Johnstone Moors

In the area south and south-east of Swimbridge the Lower Carboniferous rocks seen farther west connect in a number of major upright folds trending east or just south of east. A shales–cherts–shales succession persists. Minor folds and faults occur locally, but there is no sign of any major translocation of strata. The northern syncline closes eastwards and the upper shales of the core cease to crop out.

Two quarries north of Bydown House show the junction between basal shales of the Codden Hill Chert and overlying cherts. In the westernmost [6213 2952], shales with subordinate cherts dip at 80°/020° and pass upwards into cherts with some shales. In the easternmost [6240 2950], the cherts of the basal beds are weathered and soft, and are overlain by 12 m of thinly bedded cherts with interbedded shales.

Roadside diggings [631 296] north-west of Kerscott contain grey and buff cherts, commonly weathered and friable. An old pit [6332 2979] to the north-north-east exposes the passage upwards from shales and cherts into mainly cherts. Weathered grey and buff banded cherts have been worked near Kerscott and at the eastern foot of Pugsley's Hill [6332 2946; 6412 2919]. Two quarries on the north side of High Down are in cherts with subordinate shales of the main cherty assemblage [6474 2940; 6498 2938]. The more westerly shows contortions, a fault breccia trending 150°, and iridescent films of chalcopyrite. A quarry [6470 2904] on the southern side of the ridge revealed grey and black cherts with the mineral wavellite; it is the type locality for the mineral but the quarry is now all but obliterated.

In the southernmost belt of Lower Carboniferous rocks, a small roadside pit [6210 2834] near Dennington Cottages is in thinly bedded grey and brown banded cherts in tight refolded folds (Figure 10). Farther east a narrow pit [6295 2818] shows mainly shales in the walls, but debris suggests that the working followed a thin succession of cherts. No mappable cherts occur east of Irishborough [634 282]. Rubble Hills [640 281] is an area of disused quarries and spoil heaps marking the exploitation of a lenticular limestone; several small exposures remain of shales and siliceous shales with grey and black limestones.

Grey and brown banded cherts, commonly much weathered, have been quarried between Heddon and Oxford Down [653 290; 6580 2883]. A pit [6585 2915] on the north side of the chert ridge shows shales and silty shales passing up through siliceous shales into mainly cherts. To the east, pits in a similar stratigraphical position lie on the north [6676 2884] and south [6694 2856] sides of the ridge; the southern pit shows tightly to isoclinally folded cherts (Figure 11). An old quarry [6762 2868] near Deer Park Gate shows siltstones and micaceous shales, commonly cherty, with current bedding indicating right way up, which have yielded *Merocanites* cf. *compressus*.

Eastwards from Castle Hill, cherts cease to be separately mappable, and the northern crop of the formation shows only shales. Immediately east of the River Bray near Shallowford, three small pits within the narrow outcrop of Codden Hill Chert show shales, siliceous shales and siltstones with some fine-grained sandstone; some of the rocks are splintery and siliceous but no chert is present. Calcareous nodules and lenticles occur within these beds farther east [6902 2823]; lenses of hard grey limestone at North Aller [696 282] are up to 0.7 m thick.

A lane [7000 2809 to 7024 2788] just west of West Ford shows a number of exposures across almost the whole width of the Codden Hill Chert outcrop; the rocks comprise grey shales and silty shales, locally siliceous, with a little thin limestone. Exposures in and near the River Mole [7210 2758 to 7208 2738] also span the whole of the crop and comprise grey shales, splintery shales and siltstones with a few chert bands.

East of the River Bray the southern crop of Codden Hill Chert has been extensively pitted for limestone, the distribution of the pits suggesting two narrow parallel lenses, possibly the same limestone repeated by folding, in the west, and five lenses farther east. The old limestone quarries are generally flooded but show sporadic exposures of grey shales, silty shales and siltstones, cherty in places, with hard grey silty limestones, locally heavily veined with calcite. The positions of lenses of limestone are indicated by pitting [6745 2759 to 6877 2750; 6804 2746 to 6854 2743; 6891 2747 to 6909 2743; 6931 2744 to 6985 2737; 6994 2719 to 7019 2707; 7091 2708 to 7123 2708; 7218 2699 to 7298 2696]. A quarry [7000 2716] south-east of South Aller contains a 12 m-high face of shales with limestone beds up to 0.4 m thick, showing contortion next to an east–west crush zone (Figure 12). Similar strata crop out immediately to the east [7006 2714] and in the nearby stream [7020 2711].

The eastermost outcrop of Codden Hill Chert, beyond which Pilton Shales are faulted against Upper Carboniferous rocks, passes beneath terrace deposits on the north side of Johnstone Moors [737 271].   EAE

UPPER CARBONIFEROUS

Crackington Formation

Bickleton Wood to Venn Quarries

A quarry [5077 3115] in Bickleton Wood, where Prentice recorded *G.* cf. *carbonarium*, shows shales with sandstone beds up to 1 m thick. A sandy development within the Crackington Formation in a syncline beneath the mouth of Fremington Pill shows turbidite sandstones up to 0.5 m thick [5128 3319] near the shore of the estuary and up to 1.5 m thick [5151 3313] at the end of the old tramway track from the limestone quarry at Penhill. The surrounding, more argillaceous, succession contains similar sandstones up to 0.2 m thick [5130 3321], and such thinner sandstones are numerous in a stream valley [512 322] on the south side of Fremington.

Massive and thickly bedded micaceous turbidites, between Barnstaple and Bishop's Tawton, form the ridge on which stands Tawstock Lookout and have been displaced north-eastwards by a sinistral fault in the river valley. East of the river this sandstone ridge, which contains the large Venn Quarries [581 306] of ECC Quarries Ltd, trends slightly south of east and terminates [6009 3000] just north of Hannaford.

Quarries close to the River Taw show: fine- and medium-grained thickly bedded and massive greywackes with mica and plant fragments and traces of load casts, and with interbedded shales [5597 3080]; 12 m of fine-grained sandstones up to 1 m thick with thin interbedded shales [5640 3110]; and 6 m of fine-grained sandstones up to 0.6 m thick with thin interbedded shales [5639 3103].

Venn Quarries [581 306] (Plate 13) are cut in grey and brown fine-grained sandstones, locally micaceous and ripple marked, in beds up to 1.5 m thick but commonly around 0.3 m, separated by shales. Although the ridge containing the quarries is mappable as a sandstone-rich band within the Crackington Formation, the sandstone–shale ratio in the rocks exposed is commonly about 1:1; a low-level south bay [5815 3049] is predominantly shaly in its northern part. Strata along the same ridge are seen in a small quarry

[6036 3018] north-east of West Combe to comprise fairly thinly bedded sandstones and shales in roughly equal proportions; laneside exposures [6024 3000] immediately east of the farm show thicker bedded and massive sandstones.

Horwood to East Stowford

The sandstone packets within the Crackington Formation between Horwood and the valley of the River Taw form fairly well-defined ridges, and increase in number southwards (up the succession) towards the Bideford Formation outcrop. Towards the Taw valley (south of Tawstock) the sandstones become segregated into generally thicker packets, and again increase in number up the succession. Two small old pits east and west of the road [518 294] 200 m S of Litchardon Cross contain debris of thinly bedded brownish grey fine-grained sandstones. Farther east, near Lovacott Green, 5 m of rather blocky medium bedded sandstone reminiscent of the Bideford Formation crop out in an old quarry [5310 2830]. In a stream section south of Stonyland, grey shales dip at 30° northward [5336 2918] and grey thinly bedded fine-grained sandstone dips at 70°/355° [5336 2892]; 1 km to the east 2.5 m of pale green shaly siltstones dip at 70°/180° [5427 2884] near Collabear.

Small exposures of greenish grey fine-grained sandstones are common in the woodlands on the east side of the Taw valley, but the interbedded shales and siltstones are rarely visible. East of the Taw and south of Codden Hill, the broader sandstone features bear a debris of greyish brown weathered fine-grained micaceous sandstone. Vertical hard grey fine-grained sandstone crops out at Shilstone Cross [6051 2673], black shale dips at 70° northward [6170 2652] south-east of Cobbaton, and grey silty mudstone dips at 75°/345° at Stowford Cross [6204 2650].   BJW

Hearson to South Molton

A small stream east of Hearson runs slightly south of east along the strike, exposing thickly bedded brown fine- to medium-grained sandstones, locally micaceous, with interbedded shales [6068 2907] and, in a small pit [6149 2885] farther east, grey shales and mudstones contain fine-grained sandstones up to 0.2 m thick in which traces of current bedding show the strata to be right way up.

The predominantly shale–siltstone sequence of the lower part of the formation contains another well-developed packet of sandstones extending 3 km from south of Filleigh [664 276] to north of Hill [691 269], gently inclined and horizontal massive sandstones in quarries [6716 2734; 6742 2719] on the northern edge of the crop contrast with thinly bedded sandstones in the southern and central part [6749 2711; 6768 2709].

A quarry [719 266] between South Molton and the town's old railway station exposes mainly fine-grained and silty grey and brown sandstone in beds up to 0.5 m thick, with subordinate thinly bedded sandstone, siltstone and shaly siltstone. Beyond the River Mole to the east, another quarry [7217 2653] in the same sandstone packet shows fine-grained sandstones with interbedded shales in its face and in a small SW-trending tunnel, possibly an old explosives store. These strata are somewhat reminiscent of the Crackington Formation seen in Venn Quarries. Gortonhill Moor, Johnstone Moor and Rawstone Moor to the east are typical of Crackington Formation country.   EAE

Bideford Formation

Horwood to Chittlehampton

An old quarry near Horwood contains debris of blocky feldspathic sandstone, but an exposure [5010 2720] at its southern end reveals dark grey shales dipping at 80° southward. Stream courses and tracks in the wooded area to the south around Webbery yield many small exposures of sandstones and siltstones. The southernmost Bideford Formation sandstone runs from the western margin of the district to pass south of Somers [556 256]; it is apparently significantly thicker than those below it in the succession, and may be equivalent to the Cornborough Sandstone of the Bideford district (Edmonds and others, 1979). Roadside exposures of this sandstone [5218 2578; 5220 2575] exhibit thickly bedded brown sugary sandstone dipping at 85°/175° and at 65° southward respectively, and a strong feature [533 256] bears a dense brash of blocky pink feldspathic sandstone, as does a feature farther east [545 253].

Brown-weathered thinly bedded feldspathic sandstone is exposed in the churchyard [5295 2687] at Newton Tracey, and 2.5 m of black shales are exposed in a stream section [5225 2658]. Exposures are rare in the ground bordering the west side of the River Taw and it is possible that there are fewer sandstones in this tract.

East of the River Taw the Bideford Formation as a whole gradually thins, and the individual sandstones also appear to thin eastwards. About 2 m of thinly bedded brown feldspathic sandstone dip at 75°/175° [6032 2645] near Kewsland, and in the track leading south from this exposure grey muddy siltstone dips at 78° southward [6022 2620]. In a road cutting [6138 2579] north-east of Hawkridge, 3 m of thickly bedded blocky brown sandstone, dipping at 65°/010°, are overlain by 4 m of grey shaly silty mudstone and 2 m of Head, and 400 m to the north-east 2 m of massive, thickly bedded brown sandstone dip at 75°/175° [6153 2612]. Heywood stands on a sandstone ridge in which 2.5 m of thinly bedded muddy feldspathic fine-grained sandstone dip at 79° southward [6295 2622], and to the south an old quarry [6305 2588] reveals 4.5 m of rubbly weathered blocky thickly bedded brown feldspathic sandstone. This sandstone passes westwards into grey shales exposed in the roadside [6282 2588]. Blocky, apparently thickly bedded, brown sandstone crops out [6475 2577] near Riding Cross, dipping at 70°/015°.

Bude Formation

Alverdiscott to Shortridge

The Bude Formation outcrop in the extreme south-west of the district contains few exposures, but broad east–west ridges are formed by packets of thickly bedded sandstone. Grey siltstones with soft brown sandstone beds dip at 80°/175° in a stream section [5027 2493]; black shale dips at 80°/165° at Woodland Cross [5395 2521; 5418 2518]; and dark grey shale dips at 70° northward [5451 2506].   BJW

Bradbury Barton to Newton

In the eastern Bude Formation outcrop, buff sugary feldspathic sandstone is exposed in a stream [6662 2656], near Bradbury Barton, and medium- to fine-grained thickly bedded and massive buff feldspathic sandstone in an old pit [6701 2628]. East of the River Bray a narrow east–west pit [6872 2651] near Hill shows silty micaceous sandstone with plant remains in its north wall, and thicker bedded fine-grained micaceous sandstones overlain by grey and black shales containing nodules in its south wall. The ridge to the south-east shows debris of sugary feldspathic sandstone at its western end and outcrops of massive and thickly bedded sandstones to the east [6997 2630]. The presence of massive fine-grained sandstone on the western side of the River Mole [7209 2628] near South Molton suggests a northern limit for the Bude Formation outcrop.

Laneside exposures [7016 2490] near Ford Down show 15 m of thinly bedded and finely laminated grey silty sandstones dipping at 67° northerly. Some 30 m of greenish grey finely laminated sandy siltstones and micaceous silty sandstones dip at 75° southerly in

Furze Bray farmyard [710 249]. On the west bank of the River Mole, Blackrock Quarry [723 251] contains fairly thickly bedded fine-grained and silty sandstones, locally finely laminated, cut by steep and vertical joints trending around north-west and north-north-east. A short distance to the south, roadside exposures show flaggy silty micaceous sandstones [7238 2492] and interbedded feldspathic sandstones and argillaceous sandstones [7245 2474] dipping at 75° to 78° northerly. About 200 m NE of Grilstone, grey feldspathic sandstones and sandy siltstones dip at about 85° northward and show wedge-bedding [7323 2484], and in the River Yeo to the east vertical thickly-bedded and massive greenish grey fine-grained sandstones strike east–west [7354 2471].

A quarry [7533 2539] on the south side of the River Yeo near Waterhouse is cut in 12 m of thickly and thinly bedded fine-grained sandstones with interbedded shaly siltstones. local current bedding shows the strata to be right way up, and there are traces of spheroidal or nodular structures. Quarries east of Newtown show platy siltstones and silty shales overlain by massive fine-grained greywacke [7640 2512], micaceous siltstones with plant remains and interbedded fine-grained sandstones and shales [766 251], shales and silty shales in the walls of an east–west pit 15 m wide from which the sandy beds have been removed [766 252], and fine-grained sandstones, commonly feldspathic and sugary, in beds up to 0.8 m thick with interbedded siltstones and shales [7679 2515].

A small stream [7696 2496 to 7710 2477] bordering South Hayne Plantation cuts through interbedded sandstones, siltstones and mudstones which dip mainly at 30° to 80° southerly but locally at 70° northward. Similar exposures occur in another small stream to the east [782 246]. It is probable that in the east of the district only a narrow outcrop of Crackington Formation intervenes between the Bude Formation and the Brushford Fault.   EAE

CHAPTER 4

Pleistocene and Recent

INTRODUCTION

Boulder clay, which has been mapped to the south-west of Barnstaple, has some associated lake clays and glacial sand and gravel. Pebbly clay and sand derived from the glacial deposits has been planed off at 1st Terrace level, and terrace deposits of this age pass westwards into raised beach. The 2nd, 3rd and 4th terraces are aligned with the present drainage system, but two terraces at higher levels (5th and 6th) trend westwards from the eastern margin of the district, passing north of South Molton, to the valley of the River Bray. Recent deposits include river and estuarine alluvium, peat, and weathered material which grades into, and is indistinguishable from, the widespread Pleistocene head.

A possible Quaternary chronology is summarised in Table 7 and discussed in pp. 55–57. The stage names used offer a convenient framework for a succession of cold and warm periods, but it is worth recalling that Bristow and Cox (1973), working in East Anglia, came to the conclusion that the Hoxnian to Ipswichian stages including the Wolstonian cold period were equivalent to the Eemian; the Anglian would then, probably, correspond to the continental Saalian.

Table 7 Quaternary chronology

Epochs	Stages (with approximate dates B.P.)	Stone ages	Cultures	Deposits		Drainage patterns and landscape
Recent			Modern Iron Age Bronze Age Beaker Folk			Minor landslips
	Flandrian	Neolithic Mesolithic	Windmill Hill Cresswellian	Peat Scree Head	Alluvium and modern beach	Marine transgression with flooding of estuary
	— 10 000 —	Upper Palaeolithic				
Upper Pleistocene	Devensian — 70 000 —		Magdalenian	Head		Frost heaving
	Ipswichian		Mousterian	Pebbly clay and sand of the estuary	1st Terrace (elephant remains at Barnstaple)	Present flow direction of Fremington Pill stream induced River Yeo and Bishop's Tawton stream effect captures
	—100 000—			Boulder clay		
	Wolstonian		Levalloisian	Lake clays Glacial gravel Pebbly clay and sand of Hele and Herton	2nd Terrace 3rd Terrace 4th Terrace	Fremington Pill stream flowed east-south-east. Temporary reversal of flow of lower reaches of River Taw. Westward drainage through Bideford. Ice-impounded water of Bristol Channel escapes via Somerset levels, Chard Gap and River Axe to English Channel
	—200 000—	Lower Palaeolithic		Boulder clay		
Middle Pleistocene	Hoxnian		Acheulian Clactonian			Meltwater from Exmoor changes drainage pattern to south to north in late Anglian – early Hoxnian
	—220 000—					
	Anglian				5th Terrace 6th Terrace	Local ice-cap on Exmoor Incipient corries form Westward-flowing river passes north of South Molton to Barnstaple; possible westward drainage north of Exmoor
	—320 000—					

Glacial and fluvioglacial deposits

Boulder clay, consisting of blue-grey silty stony clay, commonly calcareous, has its broadest outcrop of about 1.6 km south-east of Combrew [5235 3222]. In this area, at Higher Gorse Claypits, lake clays within the boulder clay are worked by Brannam's Pottery, Barnstaple. Assuming a regular base to the clays, a north–south section drawn across the deposit suggests a maximum thickness of about 23 m just south of the pit. Maw (1864), who first recognised the clays as boulder clay, reported 78 ft (23.8 m) of clay in a well described as at Roundswell; Roundswell House [5361 3053] appears to be off the boulder clay outcrop, but a locality named Roundswell [543 313] lies within it.

Trenches [4982 3206] west of Fremington showed sticky red, grey and blue clay, locally silty, up to 1 m thick. Exposures farther east show 2.5 m of stony silty clay, possibly in part head [5043 3195], and 3 m of pebbly sand and silt [5079 3240].

Clampitt [5266 3286] is said to have been a pottery. An old pit remains, together with clay debris and pebbles. Fragments include a large flint and some grey limestone.

Scattered erratics in the central area of the boulder clay outcrop include sandstone, quartzite, granite, tuff, dolerite, ?quartz-porphyry and andesite (Dewey, 1910; Taylor, 1956). Pieces of oolite up to 0.4 m across in a ditch [5247 3119] were probably dumped.

Numerous small disused pits are scattered throughout Claypit Coverts [527 319]. They are now degraded, overgrown and filled with rubbish, mud and water. The deepest are said to have been dug to about 6 m. Present working is confined to a single pit, Higher Gorse Claypits, whose section showed, in 1970, lake clays within boulder clay:

	Thickness m
Soil, head and weathered boulder clay	
Thin soil and subsoil on pale grey and brown clay and silty clay containing many rock fragments and pebbles — mainly sandstone and quartzite but with some slate, vein quartz, chert, granitic rock, pebbly quartzite, dolerite and flint	4
Boulder clay	
Chocolate-coloured clay, commonly silty, with scattered small pebbles and stones as above and some lignitic wood	0.5 to 1.75
Lake clay	
Chocolate and purplish grey sticky tenacious clay, almost entirely stone-free	6
Boulder clay	
Stony clay	

Stephens (1966) reported a well-rounded gravel at the base which he equated with the raised beach of the coast. He noted a preferred pebble orientation of NNW–ESE. However, his suggested connection between the gravel and the pebbly drift of Penhill Point is neither demonstrable or likely (Edmonds, 1972a), and Kidson and Wood (1974) have shown distinct sedimentological differences between the two deposits.

The boulder clays and lake clays are not separately mappable. However, in 1972 Messrs C. H. Brannam Ltd drilled 18 shallow holes in an attempt to delineate the stone-free pottery clays. The latter may be taken to be the lake clays, and their extent is described in Chapter 8. Evidence from individual boreholes south of the present working is as follows: head on boulder clay 5.2 m, smooth clay 3.0 m, on gravel [5284 3161]; head on boulder clay 4.0 m, smooth clay 4.9 m, on gravel [5294 3159]; head on boulder clay 2.7 m, clay (probably boulder clay) 3.6 m, on gravel [5284 3153]; head on boulder clay with a few stones towards the base 4.0 m, smooth clay 2.1 m, boulder clay 0.9 m, on gravel [5290 3152]; head on boulder clay 3.7 m, boulder clay with a few stones 5.5 m, on gravel [5296 3149]. Evidence from boreholes north of the present working is as follows: head on boulder clay 3.4 m, smooth clay 5.5 m, boulder clay 0.6 m, on gravel [5297 3192]; head on boulder clay 3.4 m, smooth clay 6.1 m, on gravel [5302 3191].

The only borehole to prove workable smooth clay outside the immediate vicinity of the present working was 600 m to the north-east [5350 3201]. It penetrated head on boulder clay 4.3 m, smooth clay with a very few small stones 2.7 m, boulder clay 0.6 m, on gravel. It is noteworthy that gravel was recorded in the bottom of only those holes which penetrated smooth clays. Kidson and Wood (1974) regarded the pebbly material beneath the smooth pottery clays as fluvioglacial outwash gravel, but Edmonds (1972a) likened it to pebbly boulder clay.

Nine boreholes sunk in the area north and east of Brynsworthy House [5354 3117] proved 6.1 to 10.1 m of boulder clay without bottoming the deposit. The easternmost borehole of the survey [5415 3125], in the Roundswell area and presumably no great distance from the well noted by Maw (1864), passed through head on boulder clay 12.2 m, on smooth clay 0.3 m.

Trenches [5417 3191] on the south-western outskirts of Barnstaple showed silty and sandy clay with scattered stones and pebbles and localised deposits of sticky chocolate-purple clay with markedly fewer stones and less silt 2.0 m, overlain by silty, sandy and commonly pebbly drift 1.0 m.

The surfaces of slopes leading down to the old limestone quarry [5535 3151] at Lake are distinctly clayey, but exposures [5533 3154; 5530 3152 to 5533 3150] show clayey, silty and sandy boulder clay (Plate 10), locally more pebbly than anything now exposed in Higher Gorse Claypits and somewhat akin to the drift of Penhill.

The only undoubted glacial sand and gravel in the district is a small patch [552 322] south of the North Devon Technical College, where gravel more uniform and finer than the surrounding pebbly drift has been dug from a small (now degraded) pit.   EAE

In the north of the district, flint erratics were found as surface debris at Crock Pits [6878 4913] in an area mapped as head. Stephens (1966) recorded erratics and striated pebbles from Crock Point where, he suggested, there may once have been a claypit; any till present is concealed beneath head. The locality lies on the southern slope of a major valley (of which the northern slope has been destroyed by the sea) which is continued eastwards, via Lee Abbey, in the Valley of Rocks. Stephens (1966) envisaged Wolstonian ice pressed against the north Devon coast and temporarily diverting rivers into channels marginal to the ice sheet. One such possible channel, he suggested, was the Valley of Rocks. However, the Valley of Rocks [7045 4962], a dry valley containing head deposits, displays on its south side several spurs that appear to have been truncated, possibly by Wolstonian ice.

Plate 10 Pebbly drift at Lake, near Barnstaple
Pebbly clay and sand of the Fremington boulder clay is exposed in the walls of an old limestone quarry. (A 11846)

At Lee Bay [6937 4921], 600 m east of Crock Pits and at the seaward end of the Lee Abbey continuation of the Valley of Rocks, some 20 m of head, comprising large angular blocks of local Devonian sandstones randomly orientated in a brown silty loam matrix, overlie gravel. The gravel is perhaps 4 m thick, with small well-rounded pebbles of Devonian sandstones, Lynton Slates and quartz, together with flints, set in a matrix of yellowish brown sandy loam and sand which in places is indurated to form sandrock. Locally it is roughly stratified. Water seepages near high water mark may indicate the base of the gravel, but the section is obscured at this level by vegetation and present-day beach deposits. The most obvious interpretation, that the gravel is part of a raised beach, seems to be precluded by the absence of shell material. The gravel is probably of fluvioglacial origin. The flint erratics, present in small quantity, were presumably derived from surrounding higher land where they were possibly placed by ice; a less likely source is a cover of Cretaceous rocks, since removed.   AW

Pebbly clay and sand

Immediately west of Saltpill Duck Pond [505 331], a slight rise bordering the estuary bears a soil of brown silty clay with fragments and pebbles up to 130 mm across of sandstone, greywacke and quartzite. There are no good sections but this drift clearly equates with that 1 km to the east around Fremington Pill.

The pebbly drift of Fremington and Penhill rises to about 18 m OD. It underlies much of Fremington Camp, and on the east side of the Pill extends to cap the promontory which terminates in Penhill Point. Inland the junction with the boulder clay is arbitrary.

West of the mouth of the pill, pebbly clayey sands and silts enclose a lens of clay about 5 m × 1 μ [5111 3313] and show slight stratification [5118 3315]. Seams and lenses of clay, silt and gravelly sand occur. A poorly defined upper layer of weathered sandy pebbly clay (Dewey, 1913) was thought by Stephens (1966) to resemble the boulder clay at Higher Gorse Claypits.

East of the pill, up to 2 m of pebbly drift overlie Carboniferous rocks near Fremington Station. The railway cutting [519 335 to 522 335] north of Penhill shows up to 4 m of generally unbedded silty sands with pebbles up to 0.45 m across — mainly of sandstone, quartzite and greywacke, but with some slate, chert and igneous fragments. The total thickness may be 13 to 15 m. On the west side of the promontory 3 m of pebbly sand [5173 3403] show some stratification in the form of seams of fine sand.

On the north side of the River Taw, remnants of similar drift are exposed east of Chivenor. Silty clay contains numerous pebbles up to 0.38 m across, many of which have their long axes steeply inclined or vertical. Silty and sandy clay at one place [5104 3463] contains a small raft of rubbly slate.

Sandy pebbly drift between Hele [5435 3208] and Herton [5540 3212] is markedly less clayey than the till to the south, and surface material is closely akin to the field brash of Penhill Point.

The pebbly drifts of the estuary have been equated with the raised beach (Stephens, 1966; Kidson and Wood, 1974). However, they are unlike both the raised beach of the coast (Edmonds and others, 1979) and the silts and silty clays of the 1st Terrace. Also, although most of the pebbly drift shows flat surfaces graded to the 1st Terrace and raised beach, that between Hele and Herton rises from 30 m to 55 m OD. The pebbly deposits resemble the stonier parts of the boulder clay, such as are exposed at Lake, and have probably been derived from boulder clay by a process which removed much of the finer sediment (p. 57).

River terraces

Four river terraces between Swimbridge and the River Taw (Figure 8) slope westwards in accord with the present drainage system. The upper three (4th, 3rd and 2nd) may possibly be correlated with fluctuations of the Wolstonian ice sheet (p. 56); the lowest (1st) is graded to the raised beach and is considered to be of Ipswichian age.

About 4 km east of Swimbridge lies the watershed of Leary Moors, beyond which drainage is southerly and south-westerly via the rivers Mole and Bray to the Taw. These rivers have deposited 'normal' low-level terraces, but two high-level terraces are unrelated to existing channels; they extend westwards for about 12 km from the eastern margin of the district near the old Bishop's Nympton and Molland railway station to the River Bray near Filleigh (Figure 8). The two high-level terraces constitute the 5th and 6th terraces of the system mapped between Swimbridge and Barnstaple, comprise poorly drained silt and silty clay, and indicate the presence of an earlier westerly flowing river. They are much degraded where cut by existing rivers, but are readily distinguishable on the interfluves.

The Leary Moors watershed [657 296] rises to 114 m OD. It bears recognisable, if ill-defined, topographic flats. To the east of Aller Cross [6999 2777], and extending from the Nadrid Water to Hacche Moor, wide rush-covered clay and silty clay flats of the 6th Terrace rise to well-defined backs at about 137 m OD. The terrace back is crossed by the A361 road [7085 2688] between Aller Cross and South Molton. A similar terrace stretches east from the River Mole to the Garliford Water. Drewstone Cottage [7420 2709] stands on this feature, a wet rush-covered clay flat. Brown clays between the River Yeo and North Hayne [7721 2596] cover a terrace whose back rises eastwards to 152 m OD. A corresponding terrace between Furzehill [7794 2608] and Hilltown [7854 2627] is poorly defined.

The 5th Terrace borders the A361 road from Stag's Head [6766 2778] to Aller Cross as silty clay flats up to 275 m wide whose backs rise to 122 m OD. Beyond the Nadrid Water this terrace, slightly less well marked, trends eastwards to

Hacche Moor [716 273]. The triangle of flat land between the Hacche stream and the River Mole south-west of East Marsh [7322 2767] is covered by clays and silty clays which form a bench above the alluvium and rise to 122 m OD. Clays of this terrace have been worked in a small brickyard; the old pits [7286 2725] are now flooded but reveal bare clay on their degraded banks. East of the brickyard and beyond the River Mole this terrace is poorly defined. It occurs alongside the river and appears to extend eastwards as a narrow strip roughly following the line of the old railway, broadening in the vicinity of Whitechapel Cottage [7494 2677]. East of the small stream there, the terrace continues to the Garliford Water as a flat up to 183 m wide. A terrace of silty clay at 120 m OD stretches between the Garliford Water and the River Yeo below Veraby [7735 2688], and is also present on the south side of the river thereabouts.

Silty clay sediments of the 4th Terrace stretch alongside the valleys of the 'Acland Wood stream', the south-westerly-flowing headwaters around Harford, the headwater stream north-west of Yarnacott and the 'Swimbridge stream' between Riverton Mill [6350 3012] and Swimbridge. These terrace deposits are graded to the water-sorted gravel on a rise [552 322] south of the North Devon Technical College (p. 56); the gravel is of more uniform size than the adjoining pebbly drift and its base lies at about 55 m OD.

The lake clays at Higher Gorse Claypits (p. 56) are up to at least 6 m thick and were deposited in water whose surface stood at about 30 m OD. Between 5 and 10 km to the east, silts and silty clays of the 3rd terraces of all four streams whose 4th terraces are mentioned above grade to the level of these lake clays.　　　　　　　　　　　　　　　　EAE

At the western end of Codden Hill the 3rd Terrace of the River Taw rises at its back to about 105 m OD. It consists of clayey sand with small pebbles, predominantly of sandstone, and in places it cannot be separated from the 2nd Terrace.

Silts and clays of the 2nd Terrace occur in the neighbourhood of Acland Wood [588 322] and farther east to Harford. The terrace is graded to a well-developed 2nd Terrace feature west of Braunton (Edmonds and others, 1979). The 2nd Terrace of the River Taw is of sporadic occurrence upstream of Tawstock. It is of similar composition to the adjoining 3rd Terrace and its back rises to about 70 m OD 0.5 km south-east of Bridgetown [578 265].

The 1st Terrace is widely present as silts and clays alongside the River Taw and in the small tributary valleys to the east, and this terrace is graded to the pebbly drift of Penhill Point. Upstream of Tawstock the terrace consists of silts and clays and it is somewhat patchy.

A 19th-century brickfield [562 330], now obliterated by Summerland Street, Barnstaple, yielded elephant remains from beneath 14 ft (4.3 m) of clay in February 1844 (M. A. Arber, 1977). Specimens in the North Devon Athenaeum comprise portions of tusk and a vertebra. Teeth from two straight-tusked elephants are now housed in the British Museum (Natural History), and are recorded as having come from Pleistocene deposits of Barnstaple. One of the teeth is labelled as having been found in a brickfield at a depth of 15 ft (4.6 m). It seems probable that this was the Summerland Street site, and that all the elephant remains came from deposits of the 1st Terrace, of Ipswichian age.　　　　　　　　　　　　　　　　BJW, EAE

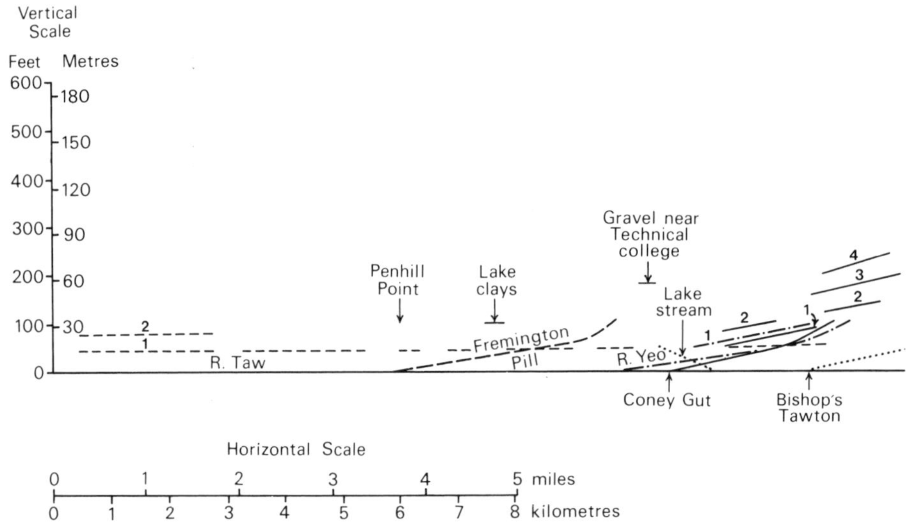

Figure 8 River profiles
in the Barnstaple district

Four terraces of the River Taw
and its tributaries are related to the
present drainage pattern. The 5th
and 6th terraces mark the course
of an early Pleistocene east–west
river, once part of the same system

Terrace deposits are rare in the north of the district, and cannot be correlated with the more extensive occurrences farther south. The valley of the River Heddon, upstream of the gorge in Hangman Grits at Voley Castle [6581 4611], contains two small terrace-like patches above the general level of the alluvium. One, a long, narrow, flat-topped spread of rounded sandstone and slate gravel with some silty sand, occurs at Bumsley [6562 4585]; the top of the terrace, here designated 1st Terrace, is about 2 m above the level of the alluvium. The other, a smaller, flat-topped spread of similar material [6572 4600], whose top is about 4.5 m above the level of the alluvium, has been designated undifferentiated terrace. The only other trace of terrace deposits in this valley is some 2 km upstream, at Holwell Castle [6707 4463] near Parracombe, where a tiny patch of rather clayey terrace, with fragments of slate and some sandstone, lies about 0.6 m above the alluvium; its relationship to the other terraces is not clear.

The East Lyn River, upstream from Brendon, is bordered by terrace-like gravels at various levels. The lowest, of small extent, occurs [7813 4818] near Ashton at about 1.2 m above stream level; behind it, and forming a well-defined bench, are gravels whose surface is about 2 m above stream level. The most extensive spreads, between Brendon and Ashton, commonly occur on the opposite side of the stream from the 2 m level and are between 3 and 3.7 m above stream level. These East Lyn gravels are all similar in composition, with well-rounded pebbles of locally derived sandstone and slate in a sandy silty matrix. AW

Alluvium

The only wide spreads of alluvium in the south of the district are those bordering the River Taw, but narrower strips extend up the valleys of tributary rivers and streams. The sediments are predominantly silts and clays, but underlying gravel is sporadically exposed in river banks. Extensive flats of estuarine mud, silt and sand (Plate 11) are exposed at low tide. On the eastern (landward) side of Penhill Point is a tract of mud and silt, 2.5 km long by up to 300 m wide, dissected by ramifying channels which fill at every high tide. The surface of the muds, however, is submerged only by exceptionally high tides and bears a cover of grasses. EAE

In the north of the district the composition of the alluvium reflects the nature of the bedrock of source areas: clay rich from slates, sandy or silty from sandstones. The valleys are generally short, narrow and steep-sided, and the alluvial deposits within them are commonly not persistent. Thus the alluvium of the River Umber at Combe Martin occurs as three separate patches in a valley whose main superficial deposits are head; it appears to merge seawards into the beach deposits of Combe Martin Beach, but relationships are obscured by buildings. A more continuous spread of alluvium occupies the Sterridge valley, which runs north to the coast near Watermouth Castle [5573 4814] and thence west-north-west, concordant with the geological strike, to become the drowned valley of Water Mouth. The only other valley containing long, continuous alluvial strips is that which runs from near Cowley Wood [6439 4525] to join the River Heddon near Hunter's Inn [6543 4830].

The valleys of the stream which flows through Shallow Ford [7136 4506], the West Lyn River, the Hoaroak Water and the Farley Water, all carry narrow alluvial spreads where they cross the Hangman Grits outcrop; little alluvium is present below the Lynton Slates – Hangman Grits boundary. Likewise the alluvial deposits of the East Lyn River occur mostly on the Hangman Grits outcrop. This lithological control of erosion and deposition has given rise to broader valleys on the sandstones than on the slates.

Storm waters have been known to effect major changes. The geological results of the 1952 north Exmoor floods were vividly described by Green (1955). Some 235 mm of rain fell on the Chains in five hours and sheet floods poured off the high ground; many of the rivers changed their courses, and there was considerable erosion and transport of materials ranging from boulders to peat, followed by deposition where valleys broadened, or their gradients lessened, or in the steep gorges of their lower reaches. Green also discussed the evidence for earlier flooding and noted the occurrence of older boulder dumps on Exmoor. The principal dumps occur on the outcrop of the Hangman Grits, where there are flood plains, and in gorges cut in Lynton Slates, where little or no flood plain is present. AW

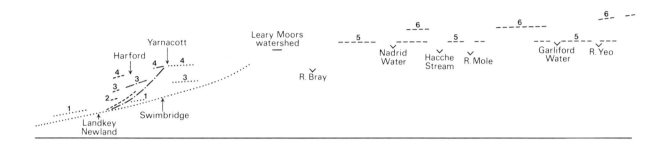

Harford
Yarnacott
Leary Moors
watershed
Nadrid
Water
Hacche
Stream
R. Mole
Garliford
Water
R. Yeo
R. Bray
Landkey
Newland
Swimbridge

Plate 11   Estuary of the River Taw near Bickington
Dissected flats of estuarine alluvium. (A 11850)

Peat

Thin peat caps several hills on Exmoor, passing at ill-defined margins into patchy peat and sandy peaty soil. A small area occurs on Winaway [719 427] and more extensive tracts, of a few square kilometres, on The Chains [731 425], Hoaroak Hill [739 427], Exe Plain [756' 423], Lanacombe [775 428] and Trout Hill [786 422]. The peat in these areas is commonly little more than 1 m thick, although on The Chains it is 2 to 2.6 m. Patchy peat is common on the high ground outside the spreads delineated on the map, and may be up to 0.5 m thick. Peat is still forming on The Chains (Curtis, 1971) but generally not elsewhere.

Farther north, thin peaty soils are present on much of the higher ground of the Hangman Grits outcrop, but patchily distributed and rarely of sufficient thickness to warrant delineation on the map. Thicker deposits of peat occur in places at the heads of valleys, where marshy ground marks the emergence of springs.   EAE, AW

Head

A mantle of weathered rock and stony clay is present over most of the southern part of the district, generally up to about 1 m thick but in places up to 3 or 4 m. It is absent on many steep slopes and on some of the high ground of Exmoor. There is little doubt that this drift cover comprises rubble generated by frost action in Devensian times, and moved downslope by solifluction during a succession of freeze–thaw cycles, together with weathering products of Recent origin. It is impossible to map these two drift deposits separately and both are classified as head.   EAE

In the Ilfracombe area the valleys of the West Wilder Brook, the East Wilder Brook and the stream which flows from Warmscombe to Hele broaden out at levels below 76 to 122 m (250 to 400 ft) OD. These lower reaches contain head deposits, with ditch and roadside exposures showing up to 2 or 3 m of stony clay, silt or sand, and impersistent narrow strips of alluvium. Cross-sections of such head-filled valleys show flat alluvium, irregular gently inclined lower slopes of head, and steeper upper slopes more or less free from drift.

The Sterridge valley is mostly flat-bottomed and filled with alluvium, but tributary valleys are filled with head. Many of the tributary streams of the River Umber at Combe Martin have head-covered valley slopes; near the sources of the streams, the head outcrop commonly widens, as near Tattiscombe [6310 4645] and at Heale [6460 4678]. In the Martinhoe Common area [670 472] tributary streams of the River Heddon have head-covered valley slopes. Near Parracombe [668 449] wider spreads of head mantle the lower parts of the long dip slope in Hangman Grits, and similar spreads occur in valleys in the vicinity of Woolhanger Common [695 460]; head in the latter area is locally peaty at surface.

In the Valley of Rocks and its westerly continuation at Lee Abbey [6980 4927] and Crock Pits [6883 4924], the gentle lower slopes and valley bottoms are filled with angular sandstone and slate debris derived from the adjacent Lynton Slates. Exposures at Wringcliff Bay [7027 4960] show scree-like head deposits extending down from about 60 m OD to beach level; the fragments within the head are progressively larger downwards and at beach level include small boulders; they are mostly angular, but some are well rounded.

Other spreads of head in the Lynton–Lynmouth area occur on valley sides rather than in valley bottoms. Head-covered valley sides characterise the outcrop of the Hangman Grits to the south and south-east of Lynton. Sharp, solid-rock dip-and-scarp features within the valleys are in places mantled by head on their lower slopes, as near Alse Burrow [7493 4471; 7488 4613] and Holcombe Burrows [7635 4421]; elsewhere they are free from head, as in Hoccombe Combe [7826 4447; 7864 4441] and Lank Combe [7857 4557]. Deposits of angular head cover the valley sides of the East Lyn River as far east as the margin of the district.

Scree

Scree is present on some of the steeper and higher slopes of the Lynton Slates and Hangman Grits outcrops, in places in sufficient amount to be delineated as a discrete deposit. Most probably it accumulated in the periglacial climate of Devensian times. The screes on the northern slopes of the Valley of Rocks, and the similar but smaller deposits [7565 5042] north-east of Countisbury Common, merge downslope into head deposits of presumed Devensian age. Generally, however, scree occurs only as isolated patches on steep slopes.

GEOMORPHOLOGY AND DRAINAGE PATTERNS

Balchin (1952; 1966) has suggested that several erosion surfaces are present in this part of Exmoor, including a 'summit surface' at 1618 to 1550 ft (493–472 m) OD, possibly representing a sub-Cretaceous peneplain; an 'Exmoor surface', preserved as a subaerial peneplain above 1250 ft (381 m) OD and representing an early Tertiary event; and a 'Lynton surface', at 1225 to 1000 ft (373 m) OD, representing a late Miocene submarine peneplain. His five lower erosion surfaces are not extensive within the present district.

Erosion surfaces, which correspond to long periods of stillstand in base level, are generally sub-horizontal and each is usually interpreted as marking the end of a cycle of erosion. Balchin (1952) favoured marine peneplanation as the cause of his high-level Exmoor surfaces. Characteristically, marine surfaces cut across geological structure and terminate against bluffs which represent old cliff lines. However, the bluffs on Exmoor are more likely to be dip-and-scarp features or other surface expressions of the solid geology. If high-level erosion surfaces are present on Exmoor, which is uncertain, the possibility that they represent Triassic planation exists. Triassic sediments occur in the nearby Vale of Porlock and just offshore from the present district, and red staining associated with faults may point to the former presence of a thin Triassic cover.

Geological control of topography and drainage is particularly evident in the north of the district. Dip-and-scarp features are common, and some valleys follow zones of faulting. Fold structures commonly have topographic expression, and in slate formations cleavage has had a greater effect on topography than has bedding.

Dip-and-scarp features generally have steep northerly scarp slopes and gentle southerly dip slopes. The latter may be related to bedding or cleavage, but are generally less steep than either. The hard, silty, indurated slates of the Morte Slates have produced a rounded dip-and-scarp topography, well seen north-west of Torrs Park at Ilfracombe. The calcareous slates of the Ilfracombe Slates have given rise to sharper features, commonly with rock outcrops on the scarps, as in the Ilfracombe area and between Ilfracombe and Combe Martin; these features are more usually related to cleavage than to bedding. Dip-and-scarp features related to overturned folds in sandy beds are visible south of Lester Cliff, and inland in the Knap Down area of Combe Martin.

Excellent dip-and-scarp topography has formed on the higher Hangman Grits between Little Hangman near Combe Martin and Butter Hill east-south-east of Parracombe. The lower Hangman Grits show fewer lithological contrasts, and features are less distinct. The NW-trending ridges of Ilkerton Ridge, Furzehill Common and Cheriton Ridge are each composed of a number of low, rounded hills separated by cols. Solid geological features run east-south-east across the ridges to produce modified and subdued dip-and-scarp topography which results in steep NE-facing and gentle SW-facing slopes bordering the valleys of the Shallowford stream, the West Lyn River, the Hoaroak Water and the Farley Water. The southern Hangman Grits – Lynton Slates junction forms a recognisable feature, except where faulted in the east. There are no dip-and-scarp features in the Lynton Slates.

It may be that the present drainage of northern Exmoor has resulted from capture of an earlier WNW-flowing system which was possibly superimposed from a once-present cover of Ilfracombe Slates. Possible remnants of the earlier system are the col at South Dean [642 482] between Trentishoe Down and East Cleave, and those east of Combe Martin between Great Hangman and Knap Down and between Holdstone Down and Stony Corner. The ancestral west-north-westerly trend survives in the East Lyn River – Valley of Rocks channel. The Valley of Rocks has a high point at the western end of Lynton village [7146 4934] from which the valley floor slopes gently away in both directions, to north-west and east. It may have served as a channel marginal to ice (Stephens, 1966) and was possibly in existence long before the Pleistocene period.

The West and East Wilder brooks, which now enter the sea at Wilder's Mouth [5190 4784], must have done so via Ilfracombe Harbour before being diverted by marine capture. Similarly the streams which rise in the Trayne Hills flowed through the presently dry valley between Hillsborough and Chambercombe before being captured and diverted north-north-eastwards by the stream which flows through Hele [534 474]. The River Umber valley at Combe Martin, the higher reaches of the River Heddon valley near Parracombe, and the East Lyn River valley near Brendon, all follow fault lines, and topographic and vegetational changes are evident on either side of the major strike faults which separate Ilfracombe Slates from Hangman Grits and Lynton Slates from Hangman Grits to the north. The River Heddon near Parracombe and its western tributary near East Middleton both flow west-north-westwards on Ilfracombe Slates, their courses determined by the strike of the incompetent argillaceous rocks, before they swing northwards through gorges in Hangman Grits. The East Lyn River and its early continuation westwards through the Valley of Rocks, past Lee Abbey and Crock Point to Woody Bay, are virtually restricted to the outcrop of the argillaceous and weak Lynton Slates.

Although a few easterly trending valleys, such as Hoccombe Water, Hoccombe Combe and Lank Combe, follow strike within the Hangman Grits, the larger valleys which cross the grits run oblique to strike. The Sterridge valley trends north across Morte Slates and Ilfracombe Slates, and the River Heddon flows north between Voley Castle and Heddon's Mouth. In contrast, several major streams crossing the Hangman Grits east of Parracombe, the Shallowford stream, the West Lyn River, the Hoakoak Water and the Farley Water, run roughly north-westwards. Some variation between westerly and north-westerly trend may reflect control by cleavage and strike in sequences of alternating slates and sandstones. The course of the Shallowford stream follows this pattern, with short westerly sections on slates interspersed with longer north-north-westerly stretches on sandstones.

The prominent WNW-trending ridge of The Chains is formed by Ilfracombe Slates whose cleavage dips steeply to the south-south-west. This steep cleavage dip contrasts with the relatively low bedding dips of the Hangman Grits to the north, and may explain the greater resistance of erosion of apparently less competent beds. Southern tributaries of the East Lyn River change course slightly to north-north-easterly as they approach and cross the Lynton Slates – Hangman Grits junction. The anticlinal disposition of the northern Hangman Grits, well seen in the coast sections of Lynmouth Bay, is reflected in ESE-striking hog's back topography in the Countisbury area, where northern tributaries of the East Lyn River run parallel to or down dip.

The coastline between Combe Martin and Foreland Point is commonly referred to as a classic locality for 'hog's back' cliffs, which typically comprise wave-truncated cliffs below a ridge shaped like a hog's back. However, the hog's back features in this area are generally asymmetrical, with a steep seaward-facing scarp slope and a less steep inland dip slope; in the case of Little Hangman and Great Hangman this interpretation in terms of dip-and-scarp features is particularly clear. Only east of Lynmouth Bay is there a nearly symmetrical hog's back topography, and even this is probably mainly a topographic expression of the anticlinal structure.

AW

PLEISTOCENE AND RECENT CHRONOLOGY

The initial Tertiary drainage pattern may have been mostly superimposed from a Cretaceous cover and ran generally south or south-east (Edmonds and others, 1969, 1975). The River Mole joined a Taw–Torridge system en route to the Exe. Evidence from the Eocene–Oligocene sediments of the Petrockstow basin (Freshney and Fenning, 1967) and from high terraces of the River Taw points to a north-north-westerly flowing ancestral river and it may be that a major reversal was effected by movements associated with the emplacement of the Lundy Granite (Edmonds, 1972a).

In early Pleistocene times a westerly flowing river ran across the present district, passing north of South Molton, to Barnstaple and the River Taw (Table 7). It ran on alluvial flats now marked by the 6th and 5th terraces as far as Stag's Head, beyond which it passed through a narrow valley cut in resistant Codden Hill Chert immediately west of Castle Hill [671 285] and thence across Leary Moors, now a watershed. Obsequent streams cut back northwards and southwards from this westerly course, along lines parallel to those of the earlier southerly drainage from Exmoor. Falling sea levels, probably during the build-up of Anglian ice, produced degradation which left earlier alluvial levels as what are now the 6th and 5th terraces. Snow accumulated on Exmoor, a local ice-cap formed, and small glaciers moved down the valleys. Corries began to be shaped; the best example is Ravens' Nest [778 409], but others occur nearby [768 411; 769 410; 781 410].

The ages of the 6th and 5th terraces are not known. The only published speculation is that of Edmonds (1972b), which implied an Anglian age. Kellaway's (1971) vision of an extensive Anglian ice sheet, with far-travelled ice abutting against local ice-caps over the whole of south-west England, is incompatible with the absence of inland erratics. Nevertheless, local ice-caps may have formed and presumably included one on Exmoor. Edmonds's (1972b) suggestion was that towards the close of Anglian times the westerly flowing river gathered the meltwaters from Exmoor to the north, which had earlier continued southwards rather as they do to-day. Such events would probably have produced considerable spreads of fluvioglacial deposits, of which no recognisable remnants remain, and it must be assumed that all have been redistributed by rivers or in head. Being of local derivation they would not have contained foreign erratics.

Towards the end of the Anglian glacial period, or early in the succeeding interglacial, meltwaters streaming south from the ice cap apparently breached the east–west divide at various times and in different places, leaving the Swimbridge – Barnstaple flow much diminished. The degree of incision of the River Bray suggests that this channel may mark the earliest breach. This accords with the Bray being the westernmost of several southerly flowing streams (Figure 8), and therefore in the position where the greatest volumes of meltwater gathered as they coursed first southwards and then westwards to the sea. Later breaches released water which carved out the channels of the River Mole and neighbouring streams, establishing the pattern of drainage to which all later terraces are related. Evidence of early westerly flows north of Exmoor (p. 56) suggests that drainage development there may have followed a similar pattern, with meltwaters running north from the ice-cap.

Hoxnian seas may have reached about 30 m OD (Orme, 1962). Mitchell (1960; 1972) and Stephens (1961; 1966) assigned the raised beach and overlying sands to this interglacial, largely on the assumption that supposed beach sediments at Fremington passed beneath the boulder clay. However this is unlikely to be so (Edmonds, 1972a; Kidson and Wood, 1974), and it now appears that although deposits of Hoxnian age were probably laid down none has survived.

Comparisons with south Ireland have led to general acceptance of the view that the boulder clay of the Barnstaple area is of Wolstonian age (Mitchell, 1960; 1972; Stephens, 1961; 1966; 1970). The Wolstonian ice advanced southwards across the Irish Sea and moved up the estuary to Barnstaple, blocking the River Taw at a point which topography and the distribution of boulder clay suggest lay between Rumsam [565 318] and Bishop's Tawton. At this stage the River Yeo, draining western Exmoor, probably flowed south through the gap at Ivy Lodge [5825 3365] to pick up the westerly flow from Landkey and continue through Rumsam to the Taw. It

also seems likely that the flow of the latter was reversed to Chapelton [580 260], whence the waters escaped westwards past Newton Tracey towards Bideford and an unknown outlet to the sea along the then southern edge of the ice sheet. Impounded water of the Bristol Channel may have escaped southwards across the Somerset Levels and through the 'Chard Gap' to the River Axe (Stephens, 1970; Edmonds and others, 1979).

Slightly warmer conditions brought about westward recession of the estuary ice and resultant deposition of boulder clay. Sandy pebbly drift between Hele [5435 3208] and Herton [5540 3212], markedly less clayey than the till to the south, probably reflects sludging and solifluction of material from the boulder clay with infiltrating meltwater carrying away the finer particles. Water-sorting produced the gravel near the North Devon Technical College which can be correlated with the 4th Terrace. The evidence suggests that in '4th Terrace times', during a slight retreat of the estuary ice, streams flowing west past Acland Wood [588 322] were joined [579 323] by waters running south-west from Goodleigh probably together with the River Yeo, still flowing through the Ivy Lodge gap. The combined waters continued westwards through the Sticklepath area.

Readvance of the Wolstonian ice seems to have followed, with falling sea level and consequent down-cutting by rivers. The ice front stood for a time at a position probably not far from Fremington. A lake formed in a basin to the east, where fairly tranquil water received the fine muds now worked for pottery clay at Higher Gorse Claypits. These clays correlate with the 3rd Terrace, and in '3rd Terrace times', as the ice front remained fairly stationary and lake clays were deposited, drainage from the east followed channels generally similar to those of '4th Terrace times'.

The ice front then apparently resumed its advance, overriding the lake basin and perhaps again reaching Barnstaple. Sea level fell and streams began to cut down through the river deposits to establish lower alluvial plains, now the 2nd Terrace.

Final melting of the ice, as the Wolstonian glaciation waned, left boulder clay which now extends westwards from the River Taw south of Barnstaple. Some 3 m or more of boulder clay overlie the lake clays.

Two river captures occurred during the Ipswichian Interglacial. The River Yeo, cutting back eastwards from Pilton [557 338], intercepted waters from western Exmoor which hitherto had flowed south through the Ivy Lodge gap, and the stream now debouching at Bishop's Tawton also cut back eastwards and east-north-eastwards to Landkey Newland, where it captured the headwaters system of the Coney Gut stream. The extensive terraces near Acland Wood border a broad valley which carries a stream much too small to have been responsible for either the erosion of the valley or the spreading of the river deposits.

The 1st Terrace is of Ipswichian age. Its profile grades to the bench of Penhill Point, but its silts and clays contrast sharply with the pebbly drift of Penhill. The latter is a mixed assemblage resembling the surface material between Hele and Herton and somewhat similar to the stonier parts of the boulder clay at Lake. It seems likely that the pebbly drift which now rests on a shore-platform mainly around the mouth of Fremington Pill and on Penhill Point has been derived from the boulder clay. Some sorting has occurred, both fluvial and solifluction processes have acted, and the sediment has not travelled far. Probably the glacial drift flowed slowly downhill, as a sludge, during which process much of the finer mud of the matrix was washed out. In some places local currents produced lenses of better-sorted material, whereas in others small rafts of country rock survived the journey. The sediments accumulated on and around an old platform in slow-moving estuary waters which planed them off at the then alluvial level. Mitchell (1960; 1972) and Stephens (1961; 1966) equated the pebbly drift of Penhill Point with the raised beach, and assigned both to the Hoxnian. Both contain derived glacial erratics, but the nature of the Penhill drift suggests a close genetic relationship with the nearby boulder clay. Also it is impossible to envisage the Penhill material being overridden by Wolstonian ice and surviving in its present form.

Topography suggests that at some time a reversed Fremington Pill stream flowed east-south-eastwards to join the River Taw via the tributary valley at Lake. Such a flow may have accompanied an early ice-induced reversal of the lowest reaches of the Taw or have been associated with meltwater marking the beginnings of the final westward retreat of the ice. However the present Fremington Pill stream can have originated only after the retreat of the ice, that is in the Ipswichian; traces of 1st Terrace occur alongside the north-north-easterly flowing stream at Fremington.

The southern limit of the Devensian ice sheet probably lay in the northern part of the Bristol Channel and adjacent areas of South Wales (Quaternary map UK, *Inst. Geol. Sci.*, 1977). Sea level began to fall towards the end of the Ipswichian as ice built up to the north. Beach deposits on the coastal platform were left above high water (Edmonds and others, 1979) and inland streams began to cut down through their alluvium, which thus became the 1st Terrace. Devensian ice advanced southwards to the northern side of the Bristol Channel, and periglacial conditions prevailed to the south of the ice sheet. To this final cold period may be attributed the solifluction which produced the widespread mantle of head, and the frost heaving which has resulted in small-scale disruptions of superficial deposits.

In Recent times the present-day alluvium has been spread over stream, river and estuary floors. Sea level has risen during the Flandrian transgression, flooding the lower reaches of the River Taw. Flandrian peat developed on high Exmoor; it is of blanket bog type (Curtis, 1971), derived mainly from sphagnum mosses and grasses, and sections commonly show a thin layer of stony head beneath the peat. Weathered debris of Recent origin merges imperceptibly into head of Devensian age.

Small landslips on the north and south slopes of the Exe valley near the eastern margin of the district are said to have formed during the great Exmoor flood of 1952.

In that part of the district north of the Exmoor watershed no detailed Quaternary chronology can be worked out. EAE

CHAPTER 5

Igneous rocks

Only two examples of igneous rock occur within the district, the tuff at the base of the Pickwell Down Sandstones (described in Chapter 2) and a dyke intrusive into Pilton Shales north of Fremington.

The dyke [5175 3365] is about 0.8 m wide, trends 290°, and has a pitted rusty brown weathered appearance. It is exposed on the foreshore of the River Taw estuary, on the west side of the Penhill Point promontory, within shales and silty shales with thin limy lenses.

The description which follows is by Dr J. R. Hawkes. The rock is autolytic kersantite lamprophyre, consisting of completely altered olivine phenocrysts set in a fine-grained matrix of oligoclase laths, biotite, secondary chlorite and carbonate (probably mainly calcite), and abundant grains of accessory ilmenite (possibly with a little intergrown magnetite); traces of pyrite are present, and some oxidation to limonite is evident. The olivine has been replaced mostly by calcite but in some cases by mineral aggregates which include quartz and chlorite. Carbonate grains in the matrix have replaced small olivine crystals, and probably also pyroxene. Some of the matrix chlorite may have been derived from these two minerals, but most occurs as pseudomorphs after biotite. Scattered throughout the rock are irregular rounded aggregates of calcite and quartz typically up to 5 mm across. Their form suggests vesicles filled by calcite and chlorite during the phase of autolytic alteration, and it is the solution of these minerals that has produced cavities in the weathered rock; the latter contains much limonite derived mainly from the autolytic chlorite.

Dr Hawkes suggests that the dyke has an affinity with the Permian lamprophyric and basic volcanic rocks to the north and east of Dartmoor. EAE

CHAPTER 6

Geological structure

GENERAL ACCOUNT

The major structure in the northern part of the district (Figure 9) is the Lynton Anticline, well picked out by the outcrop of the Hangman Grits (Edmonds and others, 1975, figs. 10 and 31). Its axis runs from Lynmouth Bay east-south-east to the Wingate [781 488] area, and the parallel Lynmouth–East Lyn Fault lies close by to the south, forming the boundary between the Lynton Slates and the Hangman Grits. North of the fold axis a pattern of smaller-scale folds and associated faults marks the southern border of the Bristol Channel trough, but to the south the general structure is much simpler. Slaty cleavage dips southwards, ranging from 30° to 40° in the Lynton Slates in areas away from major faulting, through 50° to 60° in the Hangman Grits, to 50° to 80° in the Ilfracombe Slates immediately south of the Hangman Grits outcrop; this change from north to south across the southern limb of the anticline suggests northward vergence of a major anticlinal structure. The rocks in the southern limb of the Lynton Anticline lie in the upstanding block south of the trough and in the northern limb of the great synclinorium whose main axis trends east from the coast a few kilometres north of Bude and which occupies most of central Devon.

Progressively younger formations come on southwards. Mapping has revealed some repetition by medium-scale folds, particularly north and south of the Pilton Shales outcrop. These folds range from wavelengths of a few hundreds of metres in the Stoke Rivers area [633 354] to a large syncline–anticline pair between Brayford [686 348] and North Molton [737 299] which shows wavelengths of some thousands of metres. The presence of distinctive cherts in the Lower Carboniferous has facilitated the detailed delineation, between Fremington [513 323] and South Molton [713 260 (Plate 9), of folds intermediate in scale between those of Stoke Rivers and those of Brayford – North Molton.

In general the fold axes are sub-horizontal or show gentle westerly or easterly plunges. Fold attitudes display a change from northward-overturned in the north to upright in the south. Intensity of small-scale folding is related to lithology, the less competent argillaceous formations having accommodated most of the regional movement. Attitudes of small folds reflect the major structures, with close northward-overturned folds in the Ilfracombe Slates and Morte Slates and near-upright folds in the Pilton Shales and overlying Carboniferous rocks. The Hangman Grits, Pickwell Down Sandstones and, to a lesser extent, the Baggy Sandstones, are relatively free from extensive crumpling.

No evidence has been found to support structural hypotheses which involve widespread thrusting in the Devonian rocks (Holwill and others, 1969) and in the Carboniferous rocks of the south-west of the district (Prentice, 1960a; Reading, 1965). Some major strike faults occur in the north, and minor ones locally elsewhere, and small bedding-plane slips are abundant. But there has been no great low-angle thrusting of strata.

Faults with a north-westerly trend and a large horizontal component of movement are numerous. The wrench displacements are mainly dextral; they rarely exceed 0.5 km and most are much less. These north-westerly faults are commonly linked by a set trending roughly north-eastward.

It seems probable that the main structures of the district are attributable to a single orogenic phase of late Carboniferous age. North–south compressive forces produced regional folds on easterly or east-south-easterly axes, with parasitic drag folds and with intense crumpling of the weaker shaly formations. Some minor fracturing took place in the fold hinges. Continued compression produced reverse faults in fold hinges and a regional pattern of faults in which dextral displacements on north-westerly lines were to some extent balanced by sinistral displacements on north-easterly lines. Field evidence indicates that in general the faults post-date the folds. Relaxation of pressure allowed some vertical settling movement both on lines parallel to strike and on existing wrench faults.

During the Alpine orogeny of Tertiary times there was some reactivation of fractures, those with a north-westerly trend being strongly affected (Ussher, 1913; Blyth, 1957; Dearman, 1964; Shearman, 1967; Edmonds and others, 1975), and slight movements still occur occasionally (Edmonds and others, 1968).

The structure is a simple one of mainly gentle southerly regional dips in the range 10° to 30° and local flexures. Generalised thicknesses of formations given on p. viii point to the base of the Hangman Grits lying at a depth of almost 4 km both beneath the synclinal axis east-south-east of Brayford and in the southern part of the district.

Bott and others (1958) interpreted the northward-decreasing gravity values across Exmoor as evidence of the overriding of low-density Upper Palaeozoic rocks by the Devonian of north Devon, a concept already applied to the Quantock Hills by Falcon (in discussion of Cook and Thirlaway, 1952). This conjectural break became known as the Cannington Thrust and was attributed to the Variscan orogeny. However the geophysical evidence can also be explained in terms of facies changes, perhaps as due to the presence of thick sandstone low in the Devonian and wedging out southwards (Bott and Scott, 1966; Cornwell, 1971; Donovan, 1971; Brooks and Thompson, 1973). Brooks and others (1977) attributed the Exmoor gravity gradient to thick low-density Lower Palaeozoic or Precambrian rocks whose upper surface dipped southwards beneath the Devonian. Their model showed this dip as 30°, placing the surface at a depth of about 4 km on the south-western slopes of Exmoor. They assumed a similar dip for the Devonian and Carboniferous formations. There seems to be no firm evidence for the presence of a major thrust beneath Exmoor, and the structure at depth on the southern side of the Lynton Anticline appears relatively simple — in contrast, probably, to that on the northern side, where the Brooks and others model shows none of the major faulting that is probably present.

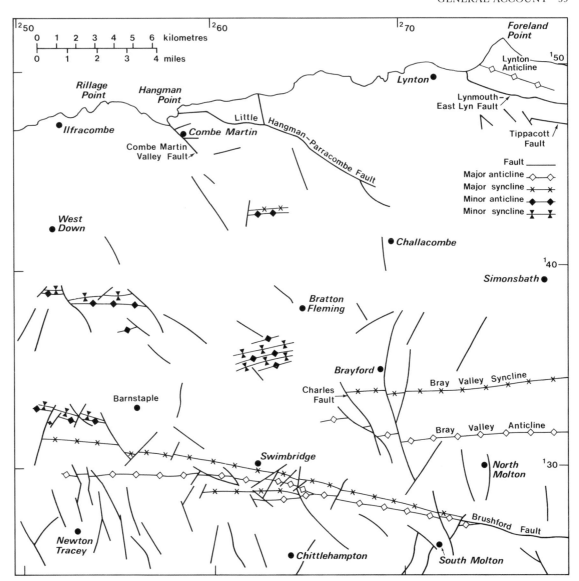

Figure 9
Main structural
elements of the
Ilfracombe and
Barnstaple
districts

In addition to the Lynmouth – East Lyn Fault, major strike faults in the north of the district are the Tippacott Fault and the Little Hangman – Parracombe Fault. Other strike faults are present in the Combe Martin area and may be traced inland where they separate contrasting lithologies. The Little Hangman – Parracombe Fault is exposed only at the eastern end of Wild Pear Beach [5822 4788]. It may be traced eastwards as a line dislocation of features, and is associated with gorges and rapids in the Hangman Grits. From the Little Hangman area to Holworthy [6869 4406] the fault separates a relatively stable but fractured block of Hangman Grits from more mobile Ilfracombe Slates; farther east it is not clear whether this boundary is a fault or a normal stratigraphic contact. At the coast the downthrow is normal, southerly and more than 180 m. Inland, the fault is responsible for local reversals of bedding dip and cleavage dip.

Several smaller strike faults are present east of Combe Martin in Ilfracombe Slates. Field relationships suggest that some may be reverse faults. One of them runs from

Coulscott to Verwill and separates Lester Slates-and-Sandstones from Combe Martin Slates over part of its course.

Shearman (1967) suggested that the NW – SE faults of north Devon were oblique-slip faults of middle Tertiary age, generally with a dextral transcurrent component. He considered that his Combe Martin Valley Fault showed a 500-ft (152.4-m) vertical displacement and a 1.5-mile (2.4-km) lateral displacement. The present work suggests a displacement of about 2 km (p. 63). Many small NW – SE faults cut the coast section west of Combe Martin, particularly in Sandy Bay [570 475]. Most dip to the south-west. Vertical and horizontal displacements are of only a few metres and many of the faults die out laterally within ten metres or so.

The coast section between Sandy Bay and Ilfracombe exposes many fold structures of around 10 m amplitude in the Ilfracombe Slates. They are tight overfolds which face north and are parasitic on the Lynton Anticline. The Ilfracombe Slates are mostly incompetent, except where sandstone or limestone beds are present, and slaty (axial planar) cleavage is well developed. This all-pervading cleavage, by far the dominant tectonic feature, tends to obscure the parasitic folds in sequences that do not have well-defined bedding. Cleavage commonly dislocates the sandstone beds, and is sufficiently intense in places to streak out *Chondrites* tubes. In less disturbed areas the cleavage in low-angle southerly-inclined normal fold limbs is commonly subparallel to bedding, but in inverted steeply-dipping fold limbs it crosses bedding at acute angles. Bedding and cleavage dips are locally not coincident in bearing but vary according to the plunges of folds. The hinge areas of folds are commonly fractured by reverse faults with various southerly inclinations. A larger fold affects the David's Stone Limestone in Sandy Bay [5690 4750] and similar structures are present farther west.

The 10-m amplitude folds carry smaller parasitic folds of similar style. These so-called 'fish-hook' structures usually occur in sequences with thin limestone bands in slates. They are tight overfolds with amplitudes of up to 0.6 m, tectonically thickened hinges and attenuated limbs; their hinges are generally not dislocated.

Of small-scale structures in the north of the district, fracture cleavage is locally present in hard, competent, thin beds near the hinge areas of folds. Joints are commonest in massive, competent beds. In relatively undisturbed areas many are parallel to the strike and more or less normal to the dip, others are at right angles to this set. Massive, competent sandstones near the faults or folded-hinges are commonly cut by joints and fracture cleavage, causing breakage into irregularly angular blocks.

Veins parallel or subparallel to bedding are locally slickensided as a result of movement in the direction of dip. Many gash veins of calcite or quartz in limestones and sandstones follow similar directions to the jointing, and their displacement may indicate the amount of bedding-plane slip; the greatest distance thus measured is 2.65 m, in the Lynton Slates of Woody Bay [6769 4903].

Dips in the upper Hangman Grits in the east of the district are mostly 30° to 60° southerly or south-south-westerly. Most of the way-up evidence points to normal successions with minor flexures, but a few steeper dips of around 75° may be associated with local small overfolds. The Wild Pear Slates and the overlying Lester Slates-and-Sandstones, between faults running north-east along the valleys east of Hollacombe [646 438] and south of Highley [679 439], show mainly slaty cleavage inclined at around 50° to slightly west of south. Dips, where evident, are at lower angles in similar directions. Cleavage within the Combe Martin Slates between the same valleys dips at 50° to 70° southerly. Farther east, beyond the fault at Highley, where no subdivision of the Ilfracombe Slates below the Kentisbury Slates is possible and all is shown as Combe Martin Slates, steep southerly-dipping cleavage prevails. Angles of 60° to 80° are common, suggesting folds only slightly overturned towards the north. A few outcrops show cleavage dipping northwards, but all recorded bedding dips are southerly. In the extreme east of the area these slates exhibit both steep south-dipping and vertical cleavage.

The Kentisbury Slates show steep south-dipping cleavage, and contain strong sandstone bands which afford evidence of dips. Around Challacombe and Challacombe reservoir, cleavage dips indicate folds only slightly overturned to the north. On North Regis Common, Broad Mead, Pink-worthy, Goat Hill and Titchcombe cleavage similarly dips steeply southerly or is vertical, trending east–west.

The Morte Slates in the west of the district display cleavage planes trending south of east. Most of these are either vertical or very steeply inclined to north or south, and suggest the presence of near-upright folds. At some places cleavage dips at lesser angles of 40° to 60°, mostly southwards but locally to the north, indicating that most overturned folds are overturned northwards.

The disposition of outcrops and surface brash, and the nature of the topography, point to the presence of an E–W-trending syncline/anticline pair east of Kentisbury Ford, with Kentisbury Slates underlying the A39 road and Morte Slates occupying the small valley [623 428] to the north between the road and Preston House.

Silty and sandy bands in the Morte Slates immediately west and south of Challacombe show steep southerly dips associated with similarly inclined cleavage planes.

The lower Pickwell Down Sandstones west of West Down dip southerly at up to 25°, and locally at up to 45°. Sandstones immediately adjoining the boundary with the Morte Slates at Bittadon dip at 85°/190°, but about 1 km to the south, in Little Silver Quarry (p. 64), southerly dips of 45° prevail. Higher beds of the formation in this western part of the district dip at up to 35° southerly.

A small E–W rise extending from Halsinger to Beara Down Cross [5195 3881] and bearing sandstone brash indicates the presence of Pickwell Down Sandstones in the crest of a small anticline south of the main outcrop. A similar but larger anticline extends 3 km east from the fault [526 382] at Beara Charter Barton (p. 64) to Higher Muddiford [558 383].

Farther east, the bottom beds of the formation between Hartnoll Barton [557 406] and Churchill Down [594 404] dip at 60° to 80° southerly in accord with the adjacent Morte Slates. To the south, between Milltown [555 389] and The Warren [595 385], the middle beds of the formation commonly dip at 45° to 60° southerly, but steeper or vertical dips prevail near the junction with overlying Upcott Slates.

The River Yeo traverses the Pickwell Down Sandstones outcrop more or less down the regional dip from Deerpark Wood [610 396] southwards through Loxhore Cott [610 383]. Exposures in the valley show gentle (20° to 30°) and steep (60°) southerly dips in the north, and moderate (45°) to very steep (80°) southerly dips in the south. Southerly dips of 45° to 60° are common in the lower beds north and north-west of Bratton Fleming.

Along the outcrop west of Bratton Fleming the presence of wrench faults trending either north-west or north-east is by displacement of the lower and upper boundaries of the Pickwell Down Sandstones. The observed movements range from 600 m east of Higher Winsham [501 391] to 100 m at [570 405].

Farther east, the Bray Valley Syncline and Anticline are major features of the structure of the Barnstaple district. They plunge gently in a direction slightly south of west, and are transected by a series of major faults. The wavelength of these folds is of the order of 2000 m. The fold limbs dip fairly steeply, and there is evidence that in places they are inverted. North and east of North Radworthy, on the northern limb of the syncline, the Pickwell Down Sandstones–Upcott Slates boundary crosses the contours in a manner suggesting a steep north-easterly overturned dip, and a probably inverted dip of 75°/010° is visible in the Pickwell Down Sandstones [7536 3468]. The southern limb hereabouts dips normally to the south-south-east, and the inversion is thought to be confined to this locality. West of the Bray Valley Fault (below) the northern limb of the anticline is locally overturned; this is suggested by inverted dips such as 70°/315° at a place [6615 3238] 0.5 km east of Stoodleigh, and by the outcrop pattern between the Charles and Bray Valley Faults, and again is thought to be a local phenomenon.

The major faults have a cumulative effect on the axis of the Bray Valley Anticline, displacing it progressively dextrally. The Charles Fault is a north-westerly dextral wrench fault incorporating a certain amount of normal downthrow to the west; the dextral displacement is up to 600 m. A subsidiary dextral wrench fault displaces the beds some 500 m and runs from near East Buckland to Goodwells Head. Its possible continuation east of the main fault crosses the River Bray near Brayley Bridge. The Bray Valley Fault was probably initiated as a dextral wrench fault, but its major component is downthrow to the west. A north-north-easterly sinistral wrench fault transects the Bray Valley Syncline in the region of Barton Wood, and likewise has a westerly downthrow.

The Upcott Slates, at the western end of their outcrop around Winsham [499 389]; show slaty cleavage vertical and striking south of east or inclined at 40° to 80° east of north, suggesting folds upright or overturned to the south. Slates exposed near the fault at Beara Charter Barton (p. 65) show cleavage vertical, strike 095° to 105°, or dipping at 70° to 85°/200° to 205°. The northern limb of the large anticline which extends east from this fault to Higher Muddiford contains slates whose cleavage is vertical or dips steeply to west of south. Slates of the southern limb are similarly disposed, with cleavage inclined west of north in one place, and it appears that the large anticline bears small-scale folds which are upright or only slightly overturned.

Farther east, exposures near Plaistow Mill [567 377] are of slates with silty and sandy bands which are faintly cross-bedded in places and show vertical and very steep southerly dips. This suggests the presence of tight and close folds, near-upright or very slightly overturned to the north. Roadside outcrops in Shirwell [597 374] point to the same pattern of folding, perhaps with some southerly overturning.

Between Chumhill [6238 3696] and Bratton Cross [6292 3665] slatey shales and silty shales lie in tight upright folds except near the junction with the Baggy Sandstones, where southerly dips of around 45° accord with attitudes in the overlying more competent formation. Exposures around Lower Haxton [6410 3666] indicate the presence of northward-overturned close folds.

Except for a possible small faulted anticline which brings Upcott Slates to the surface south of St Michael's Church, Marwood, the western part of the Baggy Sandstones outcrop appears to comprise a steeply south-dipping succession reminiscent of that on the coast at Baggy Point, with variations in the angle of dip producing slight flexures. Such a disposition is confirmed by the large faces of Plaistow Quarry [568 373] (Figure 3), where some faults are exposed but no large folds. However, a small rise 400 m south-east of Kingsheanton [553 372] carries sandstone fragments of Baggy Sandstones type. Exposures of Pilton Shales occur

between the rise and the main crop to the north, and there is little doubt that a syncline of a few hundreds of metres wavelength is present and trends east-north-east.

In and near the valley of the River Yeo east of Youlston Park, Baggy Sandstones dip steeply southwards or are vertical, suggesting the possibility of tight folds, and perhaps some overturning, near the top of the formation. To the east, exposures alongside the old Lynton–Barnstaple railway, where it adjoins the Barnstaple–Bratton Fleming road, show southerly dips of 50° to 55°.

The western outcrops of the Upcott Slates and the Baggy Sandstones have been displaced by several faults which trend to between north-west and north-north-east. The largest displacement is about 1 km (p. 65).

Between Bratton Cross [629 366] and Stoke Rivers [633 354] three ENE-trending ridges of Baggy Sandstones, separated by Pilton Shales, attest the presence of major folds with wavelengths of some hundreds of metres. Evidence of dips is sparse, but it seems probable that the large flexures are near-upright close folds and that small tight folds have developed within them, especially in the shaly sequences.

The outcrop of the Upcott Slates and Baggy Sandstones between Bratton Fleming and North Molton reveals little in the way of minor folding, and dips generally conform with the configuration of the Bray Valley Anticline and Bray Valley Syncline (p. 60). In addition to the major faults already described (p. 61), the outcrop of these formations is affected by several other faults in the area between the Bray and Barle valleys (p. 65)

There are no horizons within the Pilton Shales distinct enough to enable major folds to be picked out. Folds at the junction with the Baggy Sandstones are mentioned above, and those at the junction with the Codden Hill Chert are described below. Pilton Shales around Heanton Punchardon [504 356] and Ashford [533 354] show steep southerly, vertical and steep northerly dips, suggesting that the minor folds are close, and upright or very slightly overturned to north or south. Exposures in the Pilton Shales outcrop south of the Bray Valley Anticline show fairly steep dips, with southerly dips predominating. The sandstones mapped in this area dip predominantly southwards. Sporadic northerly dips are also fairly steep, and the pattern suggests a south-dipping succession with some upright or slightly northerly-overturned folds. Minor folding can be seen in the quarries in the Bray Valley Syncline (p. 66).

The distinctive nature of the Codden Hill Chert has permitted delineation of a number of major folds trending east or south-of-east between Fremington and South Molton Station. These large folds vary from close to open, and are upright or perhaps very slightly overturned to the north. They bear small close folds with similar attitudes, particularly in the shaly parts of the succession.

Codden Hill Chert strata between Fremington Station and Fremington are disposed in a syncline and complementary anticline trending east–west. A fault runs south-south-west along the valley east of Fremington House. East of this fault the northernmost belt of Codden Hill Chert passes beneath Penhill [5205 3315]. A single southern crop in the neighbourhood of School Road, Fremington [5131 3235], is thought to lie in the crest of an anticline. Farther east, to the outskirts of Barnstaple, the two belts of Codden Hill Chert are cut by north-easterly sinistral and north-westerly dextral faults. Exposures between them and to the south appear to be of the Crackington Formation, suggesting that the northern belt dips south-south-westwards between Pilton Shales and Crackington Formation and that the southern belt marks an anticlinal crest. The southern (anticlinal) Codden Hill Chert passes south of Barnstaple. At Lake [555 317] it contains strata dipping vertically and steeply south-south-west.

From Rumsam [568 316] to Landkey Town [591 311] a Codden Hill Chert sequence of shales overlain by cherts capped by shales with limestone lenses dips west-of-south at 30° to 65°, commonly at about 45°. Minor folds occur locally.

The predominantly shaly Codden Hill Chert of Landkey and Swimbridge Newland [600 310] shows several dips in the range 75° to 90°. A fault west of Newland House [6060 3075] trends south and south-east, and east of it a shales–cherts–shales-with-limestone succession extends to Swimbridge. Moderate dips, rarely over 45°, prevail. The Codden Hill Chert of Swimbridge lies in the northern limb of a syncline whose axis passes south of the village approximately along the Swimbridge–Hannaford road.

The ridge that runs eastward from Nottiston to Codden Hill, across the Taw valley, is anticlinal in form, with a core of chert. The small outcrop [545 293] of Lower Carboniferous rocks south of the main ridge is thought to be part of the southern limb of another anticline, downfaulted to the south by a normal strike fault. The limbs of the Codden Hill anticline are steeply dipping, and the fold opens out at the eastern end of Codden Hill to reveal Pilton Shales below the cherts in the core. The cherts in the northern limb form the southern limb of the complementary syncline south of Swimbridge and underlie the old road north of Bydown House [623 294]; limestones in the upper shales of Swimbridge are not repeated in this limb hereabouts. The ridge of Hearson Hill [605 294] and Dennington Cross [6197 2921] marks the cherts in the core of a syncline to the south; Pilton Shales in the southern limb of this fold are faulted against Crackington Formation.

Farther east, between Swimbridge and Rubble Hills [640 281], the outcrop of the Codden Hill Chert delineates two syncline–anticline pairs within a tract of country 2 km wide measured at right angles to the strike. The northern pair represent eastward continuations of the two folds noted near Swimbridge (above); the synclinal axis trends east-south-east from Indiwell [632 297] between the chert ridges of Great Hill [640 295] and Pugsley's Hill [639 293] and the twin rises of High Down [649 292]. Opposing dips evident locally attest the presence of minor folds. The anticlinal axis, although locally displaced by faults, trends east-south-east in Pilton Shales from Kerscott [632 294] to Yollacombe Plantation [642 290].

The southern fold pair shows an east–west synclinal axis lying about 600 m south of Kerscott and an anticlinal axis passing between Dinnaton [6205 2814] and Dennington Cottages [621 285] and thence through Irishborough [634 282] to Tower Farm [640 284].

The two anticlines merge into one east of Heddon Cross [649 284], and the outcrop of the Codden Hill Chert thence to its eastern extremity lies within a single syncline–anticline pair of folds. The synclinal axis trends south of east along the chert ridge of Oxford Down [664 288] to Deer Park Gate

[6772 2855], beyond which it follows a belt of Codden Hill Chert shales to Aller Cross [700 281].

The Pilton Shales between Heddon [6485 2875] and the River Mole east of Castle Hill show vertical and steep southerly dips, suggesting the presence of many small-scale folds with some northerly overturning. The major anticlinal axis runs south of east through this tract, and Codden Hill Chert shales to the south between Filleigh [661 279] and South Aller [699 273] lie in the southern limb of the anticline.

At the eastern limit of the Codden Hill Chert outcrop, north of South Molton, where the formation consists predominantly of shales in its northern (synclinal) crop and shales with lenticular limestones in its southern crop, dips are steep both within the Codden Hill Chert and in the intervening Pilton Shales of the anticlinal core. The major folds carry minor close folds, upright or slightly overturned to the north.

It is impossible to discern major folds within the Crackington and Bude formations, except where strata of the former lie within such structures picked out by Lower Carboniferous rocks. However, small-scale folds probably reflect the presence of close to open flexures, upright or slightly overturned to the north which, by anology with the folds of Lower Carboniferous rocks (above) and Bideford Formation rocks on the coast (Edmonds and others, 1979), may be assumed to have wavelengths of a few hundreds of metres.

The thick sandstones within the Crackington Formation between the River Taw to the south of Barnstaple and Hannaford, which have been worked in Venn Quarries, lie in a major syncline. Farther east, in the country stretching from Wrimstone [613 282] through Barton Cross [648 273] and Bradbury Barton [669 264] to Kingsland Barton [699 259], scattered exposures with mainly steep opposing dips suggest the presence of close folds either upright or slightly overturned to the north. Similar attitudes characterise the area between South Molton and Bish Mill [743 254], with most exposures being in roadsides and in the valleys of the Rivers Mole and Yeo.

The general pattern of near-upright close folds within Upper Carboniferous rocks persists to the eastern edge of the district, except that where strong thick-bedded sandstones are well developed the folds are open.

The distribution and observed dips of the sandstone ridges within the Upper Carboniferous outcrop in the southern part of the district around Newton Tracey and Chittlehampton suggest the presence there of a set of upright, mostly open, anticlines and synclines, some with steep to vertical northern limbs. The structure is most clearly demonstrable within the outcrop of the Bideford Formation, with its high proportion of sandstone beds. The outcrop of the whole Upper Carboniferous sequence in this area is transected by a set of north-easterly sinistral and north-westerly dextral wrench faults, again strikingly picked out by the Bideford Formation outcrop pattern, and there are rare instances of north-easterly dextral and north-westerly sinistral wrench-faults.

EAE, AW, BJW

DETAILS

Lynton Slates, Hangman Grits and Ilfracombe Slates

The hinge of the Lynton Anticline crosses the coastline approximately at Blackhead [745 502], between the two prominent headlands of Lower Blackhead and Upper Blackhead; the general dip thereabouts is about 40°E. The strata are folded into an asymmetrical anticline whose southern limb dips at about 35°SE or ESE and northern limb at 40° to 70°NE or ENE; the easterly component of dip possibly reflects the east-south-easterly plunge of the major structure. Inland exposure is poor, but the hinge roughly coincides with the ESE-trending watershed of Butter Hill, Countisbury Common, Kipscombe Hill and Wingate. The broadly anticlinal strata of the hinge area are intensely disturbed by small-scale folding and faulting. Locally the dip is variable, owing to the presence of minor flexures and sigmoidal folds. Between the Lynmouth–East Lyn Fault and Blackhead, small-scale folds mirror the large-scale structure; southern limbs dip relatively gently to the south and are commonly fractured by reverse faults, whereas northern limbs are near vertical. Between Blackhead and Foreland Point there are abundant small-scale folds and inversions, but sedimentary structures and bedding–cleavage relationships show that in general the beds young towards the north-east.

East of Foreland Point the beds dip generally to the north or north-east at various angles, but small-scale anticlines and synclines and monoclinal flexures are present. It seems probable that the small-scale folding is intimately associated with faulting, and may be related to drag effects from nearby large faults.

The Lynmouth–East Lyn Fault affects the crestal region of the Lynton Anticline. It crosses the coast at Ninney Well [7350 4957], where it is near vertical and sharply defined. At beach level below Ninney Well it trends WNW. From Ninney Well it trends southwards to Wind Hill [7356 4931], whence it strikes ESE to beyond Leeford [770 482] and separates Lynton Slates from Hangman Grits. The diversion may be a branch fault, since a projection of the main line west-north-westwards from Wind Hill joins faulting of similar trend in red-stained Lynton Slates near Point Perilous [7291 4960]. In the east the fault traverses the deeply incised valley of the East Lyn River and veers southward where it crosses that river at right angles near Wilsham [7533 4848]. Observations and calculations throughout its inland course, and particularly near Wilsham, suggest that the fault plane dips southwards at about 45°. The fracture is a fairly steeply inclined reverse fault, similar in style to the small reverse faults which dislocate the hinge areas of many minor folds in the north-east of the district. In the Lynmouth–Countisbury area Hangman Grits resembling the Sherrycombe Formation or parts of the Rawn's Formation (p. 24) have been brought into contact with strata low in the Lynton Slates, suggesting a throw of 1800 to 2000 m or possibly more. Farther east, grits of Trentishoe Formation type are faulted against higher Lynton Slates and the throw is less, probably about 1500 m at the edge of the district. Ussher (1889) traced the fault to Brockwell near Wootton Courtenay, which indicates a total known length of about 20 km. The Lynmouth–East Lyn Fault shows features (p. 58) consistent with a fracture active locally in Lower Devonian times. Clasts of older Lynton Slates appear in younger Lynton Slates, the source area lying close to the north. The change in structural style from south to north across the ESE-trending fault line may reflect a different sub-Hangman Grits floor on either side of the fault, which may in turn contribute an effect to the Exmoor gravity gradient.

North of a line from Duty Point [694 497] to Myrtleberry Cleave [740 486], and within the area of outcrop of the Lynton Slates, bedding dips diminish from south to north towards the crest of the Lynton Anticline. Inclinations are 17° to 29° in the Lee Bay area [692 492] and locally 6° to 15° between Wringcliff Bay [702 496] and Hollerday Hill [712 499]. North of Wringcliff Bay, on the northern slopes of the Valley of Rocks, Lynton Slates inclined at about 7°E or ESE may be aligned with the plunge of the anticline. However, north of an ESE-trending line through Ruddy Ball [715 501] and Watersmeet [743 486], including the coastal exposures around Lynmouth, the Lynton Slates are more disturbed and steeper dipping, owing to the proximity of the Lynmouth–East Lyn Fault.

Several minor NW-trending faults affect the Lynton Slates–Hangman Grits boundary between Cheriton [742 470] and Tippacott [768 470]. A strike fault, the Tippacott Fault, trends E–W between Tippacott [7685 4713] and the Badgworthy Water valley [7890 4715] just east of the district. Its presence is inferred from the absence of the usual strong feature at the Lynton Slates–Hangman Grits boundary, coupled with strong red-staining of the ground and the sharp juxtaposition of slate and sandstone fragments as surface debris. The downthrow is southerly and unlikely to exceed 50 m.

A fault trends SSE from the coast near The Mare and Colt rocks [6240 4860] to Broadmoor [6263 4721], where it is truncated by the Little Hangman–Parracombe Fault. It displaces features in the Hangman Grits of Holdstone and Trentishoe downs, follows a valley marked by springs and boggy ground, and throws down eastward perhaps 25 m.

The Hangman Grits of the gently southerly-inclined southern limb of the Lynton Anticline strike E–W in the Little Hangman area and ESE in the Parracombe area, subparallel to the Little Hangman–Parracombe Fault. The ESE strike is maintained eastwards to the Farley Water. At Wild Pear Beach [5822 4788] the Little Hangman–Parracombe Fault dips at about 80° southward. On the north side lie relatively undisturbed, competent Hangman Grits; on the south side are Little Hangman Formation beds and Wild Pear Slates intensely crumpled into tight N-facing, almost isoclinal, overfolds with wavelengths of about 4 m. The tectonic thickening of the incompetent beds and their structural complexity, which imply decollement or detachment of the incompetent from underlying competent beds during major folding, make the fault's throw impossible to measure.

The Hangman Grits at the summit of Little Hangman dip southwards at up to 40°, and in the cliff lower beds of the same formation dip at 20° to 25°. This change of dip in 200-m-high cliffs is probably due to tectonic causes. On the north side of the Little Hangman–Parracombe Fault, within a zone perhaps 80 m wide affecting only the highest Hangman Grits, small-scale asymmetrical folds are present above low-angle reverse faults. The reverse faults dip at about 25°/175°; above them the northerly-inclined limbs of the folds dip at 60° to 90°/010°, and the southerly inclined limbs dip at 30° to 35°/190°. The strata are competent but strongly cleaved; in the S-dipping fold limbs cleavage is inclined at 60°/180°.

The presence inland of tight overfolds, similar to those in the Little Hangman–Parracombe Fault, can be deduced from the irregular disposition of modified dip and scarp features south of Girt Down.

Exposures in Ruckham Combe [726 438 to 727 431] in the top beds of the Hangman Grits show beds and cleavage dipping southwards, the latter at the steeper angles. Similiar dispositions were noted along Hoaroak Water [743 439 to 747 431]; in one place [744 436] the disturbed strike of vertical beds suggests the presence of a fault. Farther east, Hoccombe Water [774 435 to 785 434] traverses scattered outcrops of S-dipping sandstones and flows roughly parallel to E–W dip and scarp features.

The Combe Martin Valley Fault crosses the coast at Combe Martin beach, where it throws down Combe Martin Beach Limestone of the Combe Martin Slates to the south-west against Lester Slates-and-Sandstones, all the strata being strongly folded. The fault plane is not exposed. The fracture appears to displace the Lester Slates-and-Sandstones–Combe Martin Slates junction by about 2 km dextrally. It follows the valley of the River Umber from Combe Martin beach to Brookside Nurseries [5895 4602] and possibly as far south-east as Umber Lodge [5913 4590], beyond which the valley divides into three with only the smallest in line with the fault. The junction between the Combe Martin Slates and the Kentisbury Slates shows no displacement and it is assumed that the fault dies out at the south-eastern end of Combe Martin village. A fault trending SSE from Voley [640 458] to Blackmoor Gate [650 431] displaces Ilfracombe Slates dextrally and downthrows to the west.

In the valley north-east of Hollacombe silty slates of the Lester Slates-and-Sandstones strike 090° and 110° on the north and south sides respectively of an E–W fault [6486 4402]. In and near the old railway cutting [6620 4406] at Rowley Cross bedding and cleavage dip southwards, the latter at the steeper angles. Exposures of the same formation in the valley west of Twineford [6742 4366] and in lanesides near the farm show cleavage dips of 40° to 85°S and 50° to 60°NNE.

The dip of 35°/220° noted in Combe Martin Slates east of Highley (p. 30) relates to strata in which cleavage planes are inclined at 50°/190°. Cleavage in this formation generally dips at moderate to steep angles southwards but locally is inclined to the north, for example 35° to 45°/350° [6924 4322] in a small valley north of Challacombe Common.

Exposures of Kentisbury Slates, silty slates with sandy pods, on the north side of Mattocks Down show slickensided steep or vertical joints trending ESE [6070 4401]. Beds near the top of the formation at Patchole are crossed by two fault-controlled valleys, those of the River Yeo south-east of Bugford [600 430] and its tributary stream flowing through Kentisbury Ford [618 426]. In an old quarry [6074 4235] silty and sandy slates dip at 45°/185° and 65°/020° in the limbs of a syncline. Greatgate Quarry [613 425] shows slates with sandstone bands up to 0.3 m thick dipping at 20°/210° and shown by cross-bedding to be right way up.

An old quarry [6435 4373] in Kentisbury Slates on the north-east side of Kentisbury Down contains some slaty sandstone dipping at 50°/180°, whose ripple marks suggest possible inversion. In the same quarry, massive fine-grained sandstone 1 m, overlain by 2 m of silty and sandy slates with pods, lenses and bands of fine-grained sandstone, dips at 65°/170° to 180°; steepening of dip just where the strata pass below ground suggests that the beds may be in the inverted limb of a syncline overturned to the north.

Slates and sandstones of the same formation are exposed [6510 4235] east of Wistlandpound in what appears to have been a rough road to the reservoir; adjacent dips of 60°/165° and 75°/030° possibly mark the limbs of a close upright syncline. On Rowley Down exposures near Little Rowley, in which southerly dips of 20° to 30° (bedding) and of 60° to 75° (cleavage) are associated, presumably lie in the normal limbs of overturned folds [6551 4306; 6555 4300; 6553 4296].

In the upper reaches of Yarbury Combe, dips of 50° to 80° southward [7099 4172] and 45° to 50° northward [7124 4198] suggest the presence of open folds. Slates with sandstones in The Chains Valley [747 418], near the eastern end of the Kentisbury Slates outcrop, dip at 45° to 80° southward in folds overturned to the north.

AW, EAE

Morte Slates

Slates near Twitchen Cross [509 437] show cleavage vertical or dipping steeply to just east of north or just west of south, indicating folds either upright or very slightly overturned to south-south-west or north-north-east. A similar pattern is evident in several localities, as along the valley east of Higher Aylescott [5255 4164 to 5380 4276].

In a pit [5697 4170] near Indicott sandy lenses within slates indicate an anticline whose northern limb is vertical, strike 085°, and whose southern limb dips at 75°/180°. Sandstones, siltstones and slates dipping at 55° to 90°/185° in a small pit [5875 4245] west of Ford show slight undulation of the beds, and are cut by a bedding-plane fault containing 0.1 m of red clayey gouge.

Upright folds at Clifton [5975 4139] have limbs dipping at about 80° to both north and south.

Slates and silty slates in the roadside at Rock Cottage [6029 4102] show slight contortion of the cleavage, generally inclined at about 70°/170°. In Lock Lane Quarry [6279 4143] slates with fine-grained sandstones are cut by vertical joints trending 175°. Similar steeply inclined joints trend about 200° through slates with silty pods and lenses in an old quarry [6318 4062] at Besshill. Exposures

in cuttings of the old Lynton–Barnstaple railway show traces of ripple marks on sandy and silty lenses within slates [6450 4076], and near vertical joints trending just east of north [6449 4061].

Slates and silty slates with sandy bands in a small quarry [6573 4046] 0.65 km SSE of Stowford are cut by S-trending vertical joints. Another old pit [6632 4119] shows no outcrop but contains blocks of quartz breccia resembling a fault rock.

Between Bratton Down and the eastern boundary of the district in the vicinity of the Barle valley, cleavage and dip within the Morte Slates are both inclined at high angles towards the south, or commonly slightly east of south. A pattern of near-upright major folds is suggested, and any inverted strata which have been observed indicate northward overturning. Small-scale folds are visible in places; an open upright anticline trends at 075° at Leworthy Mill [6725 3830] and has a wavelength of about 150 m, and a similar fold trends 100° at Wallover Down Gate [6895 3937]. The axial region of a syncline pitching gently to 275° is seen in a stream bed at Little Melcombe [7183 3876]. Many minor folds can be seen in the well-exposed area south-west of Simonsbath, in the steep-sided valleys around Horcombe and Burcombe, notably a set of E–W-trending northerly-overturned near-isoclinal folds seen in quarries [748 388]; these folds have wavelengths of 10 to 15 m, and in Drybridge Combe [759 387] a syncline with a wavelength of 120 m trends 285°, both limbs dipping at 75°. Farther south exposure is poor, but the axial region of an anticline can be seen in a tributary [765 372] of Kinsford Water, trending E–W, the northern limb dipping at 80°, the southern at 70°.

EAE, BJW

Pickwell Down Sandstones

Extensive exposures in the Pickwell Down Sandstones near the old railway at Fox Hunter's Inn [508 419] show sandstones dipping at various angles but in all cases around southwards or slightly west of south, indicating southerly younging of strata which show undulation of the beds but are otherwise not folded. Outcrops of tuff at the base of the formation prove the presence of a north-westerly wrench fault with a dextral displacement of almost 300 m in a stream valley near the Fox Hunter's Inn. About 4.5 km to the east-south-east, in Little Silver Quarry [549 403], strata dipping generally about 45° southward show some evidence of movement on bedding planes which has produced thin clayey silty fault breccias. To the north-north-west, the base of the formation has been displaced 550 m dextrally by a north-westerly fault running along the valley at Bittadon, and a north-westerly fault through Honeywell [570 405] has a dextral displacement of 100 m.

Adits at Fullabrook Mine [5155 3985] (p. 74) follow manganese-rich breccia bands, probably in fractures, trending 015°. The top of the Pickwell Down Sandstones has been displaced by a north-westerly fault east of Higher Winsham [501 391] (600 m dextral), a north-easterly fault west of Halsinger [513 389] (300 m sinistral), the northerly part of a predominantly north-westerly fault north of Beara Charter Barton [524 384] (300 m dextral) and a north-easterly fault east of Middle Marwood [538 388] (200 m sinistral).

Workings [571 389] at Viveham Mine (p. 74) appear to trend about 320°, and the presence of fragments of hematitic fault breccia suggests that adits may have been driven along fracture zones. Shirwell Mine (p. 74) includes two old workings [607 383; 607 377] 377] which appear to trend around north-west, possibly following fractures. Roadside exposures [6081 3755] near the more southerly are cut by a small E–W strike fault, and a small quarry [6094 3777] contains sandstones cut by prominent joints inclined steeply between 280° and 310°. Most of the dips thereabouts are around 45° to 80° southward, but thin-bedded shales and sandstones in a roadside [6094 3731] dip at 55°N, suggesting the presence of near-upright open and close folds with E–W axes. An old quarry [6103 3813] 150 m S of Loxhore Cott shows sandstones dipping at 80°/190° with interbedded shales showing traces of vertical slaty cleavage, suggesting the presence there of upright close folds. A lane [6147 3764 to 6156 3764] at Lower Loxhore is floored by sandstones with shaly partings dipping steeply to north and south, indicating the same style of folding.

North and north-west of Bratton Fleming the base of the Pickwell Down Sandstones is cut by two north-westerly faults; that in Smythapark Wood [638 385] shows a dextral displacement of 600 m, and that near Beara Manor [645 383] a dextral displacement of 250 m.

Massive and thick-bedded fine-grained sandstones [6418 3827] alongside Button Hill, Bratton Fleming, dip at 60°/180° and are cut by small strike faults. The junction with overlying Upcott Slates is exposed to the south in a lane [6426 3677] at West Haxton. Sandstones in nearby Haxton Down Lane dip at 50°/360° and 55°/175° [6457 3688 to 6468 3690], indicating the presence of open upright folds.

Southerly dips of 60° to 85° predominate in the outcrop of the Pickwell Down Sandstones between Bratton Fleming and Long Holcombe, but the axial region of a minor open upright anticline is visible in Wind Lane [6961 3596] trending at 100°, limbs dipping at 70°/010° and 80°/190°. Two prominent sets of joints are seen in Berryhill Quarry [6697 3717], inclined at 20°/080° and 90°/210°, and other less consistent joints are commonly seen in the more massive sandstones. A NW-trending wrench fault has displaced the base of the formation dextrally for 500 m in the region of Berry Wood [677 375], and a NE-trending wrench fault has moved the base sinistrally for about 50 m near Gratton [685 373]. A wrench fault trending east of north through the area of East Down Wood has displaced the base of the formation 550 m dextrally [704 365]; it is possible that this fault also downthrows normally to the west. The base of the Pickwell Down Sandstones is displaced 950 m by a NW-trending dextral wrench fault on Western Common [723 363], and farther east [733 356] a NE-trending sinistral wrench fault displaces the whole formation by about 150 m. Two minor wrench faults displace the top of the formation by a few tens of metres south of Fyldon Common.

The wide outcrop of the Pickwell Down Sandstones in the eastern part of the district, between North Molton Ridge and Kerswell Farm, yields fairly steep southerly dips, and few minor folds are found within the Bray Valley Anticline.   EAE, BJW

Upcott Slates and Baggy Sandstones

Available structural details of the narrow outcrop of Upcott Slates between Winsham and Haxton comprise the attitudes of cleavage planes, which are summarised above (p. 61).

A disused quarry [506 378] in the top beds of the Baggy Sandstones near Boode shows thinly bedded, thickly bedded and massive sandstones, nearly horizontal or with gentle westerly dips. Exposures at the north end show brecciated sandstone between relatively undisturbed beds, and the north-west face also contains traces of near-horizontal fracture. The top of the Baggy Sandstones is displaced by north-north-easterly faults near and east of Boode, and the large NW–SE fault (p. 51) which runs from Beara Charter Barton to Whitehall cuts across the full width of outcrop. Dips at Whitehall and in the west end of Lee Wood are 60°/175° to 180° [5339 3748], 40°/350° to 360° [5349 3741] and 60°/190° [5363 3743], indicating open/close folds trending approximately E–W. Beds in a quarry [5500 3737] near Guineaford dip at 60° to 75°/175° to 180°.

South-dipping strata in Plaistow Quarry [568 373] are cut by a small vertical E–W fault [5685 3727] which is also visible, steeply inclined to the south, in the roadside [5665 3728] immediately south of the quarry entrance gates. A second fault in the face, a few metres north of the first, dips steeply northward.

A large NW–SE fault runs along the valley of the Bradiford Water from Milltown [555 389] to Plaistow Mill [567 377], and continues to displace the Baggy Sandstones outcrop dextrally by about 100 m in the neighbourhood of Sloley Barton [573 372]. A branch of this fault trends ESE to pass south-west of Youlston [586 375],

where the formation shows a dextral displacement of about 1 km. Near Shirwell Cross [590 370] this major fault truncates a north-easterly fracture along which the Baggy Sandstones show a sinistral displacement of 300 m.

Medium-grained thickly bedded sandstones dip at 80°/340° [6097 3662] near the junction of the Baggy Sandstones with the Pilton Shales. This confirms the general indications of folds at the top of the formation, where a succession containing thickly bedded sandstones is overlain by a less competent one predominantly of shales. Thin sandstones dipping at 40°/330° [6329 3640] in woodland 1 km N of Stoke Rivers probably lie near the east end of the large anticline marked by the Baggy Sandstones ridge to the west-south-west. Thinly bedded sandstones [6350 3496] in a lane south of the village dip at 75°/030° near the junction with the Pilton Shales.

A sinistral NE-trending wrench fault displaces the Upcott Slates and Baggy Sandstones by 100 to 200 m between Nadrid Water [703 288] and Flitton Green Cross [717 310], and a similar fault displaces the top of the Baggy Sandstones by about 400 m nearby to the south-east [711 293]. Farther east, a fault trending west of north displaces the top of the Baggy Sandstones and sandstone beds within the Pilton Shales by up to 200 m dextrally. A wrench fault trending NW–SE from near Oakford Copse [729 305] is unusual in having a sinistral throw, of up to 500 m. A typically dextral NW–SE wrench fault transects Lambscombe Hill [765 293].

                                               EAE, BJW

Pilton Shales

Closely adjacent dips recorded in Pilton Shales in Ladywell Wood [around 5171 3584] indicate the presence of tight folds slightly overturned to the south. Strata in the roadside [5237 3542] at West Ashford are disposed in near-upright folds with limbs dipping at 65°S and 70°N. Similar exposures [5344 3519] near Ashford show a syncline with limbs inclined at 60°/190° and 55°/020°. Roadsides [5418 3556 to 5467 3521] near Upcott show fold limbs dipping at 65° to 75°/340° to 355° and at 65° to 80°/165° to 175°.

On the estuary shore between Chivenor and Strand House the strata dip steeply to just east of north and west of south, or are vertical. A syncline [5195 3494] has limbs inclined steeply to 020° and 200°. Just west of Strand House a fault trends 070° through disturbed shales. On the shore 500 m W of the mouth of Fremington Pill an E–W syncline is overturned to the south. Several other small folds are exposed on the shore [5171 3395 to 5173 3356] north of Fremington Station; the limbs dip at 55° to 80° to within 15° of north or south.

An old quarry [5598 3508] south-west of Playford Mill (Figure 4) shows strata inclined at 45° to 60° southward and cut by a S-dipping fracture, probably a reverse fault, of very small throw. In another old pit [5620 3512] south-east of the mill (Figure 5) a small tight anticline overturned to the north is faulted along its inverted limb. Laneside and roadside sections in the Barnstaple area show dips of 85°/010° and 50°/200° [5552 3432], 85°/180° and 80°/360° [5655 3492], and 60° to 80°/185° and 75°/015° [5669 3392 to 5698 3399].

Dips near Hartpiece point to the presence there of close upright folds whose axes trend just north of east [5716 3594; 5744 3610]. A narrow cutting running south-south-west from Parkhill, probably an old packhorse trail now replaced by the nearby road, shows upright folds trending 070° [5839 3610]. An old quarry [5867 3507] just south of Coxleigh Barton shows shales and silty shales dipping at 75° to 80° southward and cut by two faults trending about 065°. A road cutting [599 357] north-east of Riversmead is in steeply S-dipping beds cut by small steep or vertical strike faults. Shales and siltstones at Chelfham show folds with E–W axes cut by strike faults [6118 3572]. Shales with thin sandstones in Horridge Wood, west of Stoke Rivers, dip generally southward but are locally disposed in an anticline slightly overturned to the south [6262 3561].

Steep southerly dips, with some vertical and steep northerly dips, prevail between Barnstaple and Landkey. A farm lane [5814 3311] at Maidenford traverses shales dipping at 75°/190° and cut by an E–W fault. A lane [597 341 to 599 343] at Goodleigh shows fold limbs dipping at 55° to 80°/010° to 025° and 75°/195° to 200°. At Combe, shales and sandstones are cut by a fault marked by white clayey gouge trending 040° [5937 3341].

Shaly exposures [5764 3152] at the side of the A361 road near Whiddon are cut by an E–W strike fault marked by a 20-mm thickness of clay gouge. Dips at Yarnacott [6226 3064] of 75°/360° and 75°/190° point to the presence of tight/close folds, and steep dips to north and south were recorded also in the stream valley to the south-east [6266 3017]. In the Bray Valley Syncline north of Charles, an anticline and syncline trending 070° with wavelengths of 40 m are exposed in one quarry [687 340], and in another [6870 3365] an anticline trending 085° is seen with its northern limb dipping at 35°, and its southern limb at 55°.

An anticline alongside the River Bray north of Shallowford has limbs inclined at 55°/045° and 75°/190° [6836 2879]. At the south-east end of the old Castle Hill railway tunnel [6912 2878] a syncline axis trends 095°; the fold limbs are near vertical and gently N-dipping. Shales at Bicknor [7412 2744] dip northward at various angles greater than 45° and show minor faults coincident with the bedding planes.   EAE, BJW

Lower Carboniferous

Codden Hill Chert underlying Fremington Station [516 333] dips at 45° to 50°/200° beneath Crackington Formation in the northern limb of a syncline. Shales with thin cherts exposed [5105 3310] on the foreshore west of Fremington Pill show much contortion and overfolding and are cut by several small faults which trend east or slightly south of east, roughly with the strike. Similar disturbed beds crop out alongside Fremington Pill [5136 3303] and on the southern outskirts of Fremington [5131 3235]. The same rocks recur in the southern limb of a complementary anticline beneath the drifts of Fremington Camp, cropping out in an old pit [5050 3284] 0.5 km to the west. Shales, cherts and limestones in a quarry [5226 3246] at Muddlebridge show small-scale contortion. An old limestone quarry [5427 3227] on the edge of Barnstaple is cut in thin cherts and limestones, generally inclined at about 40° southward but locally disposed in small open folds with E–W axes. A flooded limestone quarry [554 315] at Lake shows shales and siltstones at water-level that Prentice (1960a) thought to be Pilton Shales associated with a thrust; it seems more likely that they belong to the Codden Hill Chert which is folded but not thrust.

A small chert quarry [5789 3125] west of Hill Farm shows small faults, vertical or inclined steeply southwards, trending 200° with the strike of the beds. At the western end of an old limestone quarry [581 309] to the south-east an ESE-trending syncline is well

Plate 12   Old limestone quarry at Venn
Lower Carboniferous limestones, shales, siltstones and cherts are disposed in a syncline trending west-north west. (A 11828)

displayed, its limbs dipping at 60°SSW and 60°NNE (Plate 12). An old limestone quarry [6060 3040], in line with the larger Marsh's Quarries, contains strata disposed in a syncline whose limbs dip at 40°/015° and 35°/240°, attesting the presence of open folds in strata near the top of the Codden Hill Chert.

A small chert quarry [6155 3028] (Figure 6) shows beds dipping at 35° to 40°SSW, traversed by seven faults approximately coincident with bedding and several smaller fractures splaying off these seven. Strata between and adjoining the faults have been crumpled, and some thrusting has occurred. Similar rocks in a pit nearby [6166 3024] are cut by several small bedding-plane faults. In a larger quarry [6201 3013] (Figure 7) at Swimbridge these strata dip at about 60°SSW and are cut by three S-dipping faults at the southern end of the face. Beds near the faults, at the southern end of the quarry, have been folded, crumpled, broken and overturned. Fairly uniformly inclined strata pass southwards into a tight northerly-overturned syncline broken by multiple small fractures and faulted on its southern side. Beyond the fault, rucked strata have been thrust northward.

Laneside exposures on the eastern outskirts of Swimbridge show folded and faulted strata with some inverted beds dipping steeply southward [6218 3005 to 6225 3006], a minor fold apparently trending NNW with no inversion [6228 3007], and a minor fold trending NNE and picked out in the floor of a lane by vertical cherts swinging through strikes of 020°–140°–200° [6236 3006].

The chert ridge between Nottiston and Codden Hill is transected by several north-westerly dextral wrench faults; the one through Eastacombe displaces the beds about 40 m, and the postulated fault beneath the Taw has a displacement of 100 to 150 m.

Siltstones, sandstones and shales with limy bands at Smalldon Farm [6273 2977] dip at 70°/130°, striking atypically in line with a small fault. A small chert quarry [6332 2946] near Kerscott contains cherts dipping at 50° to 60°/200° with minor contortions of the beds and with some small strike faults. A quarry [6474 2941] on High Down exhibits cherts with shales dipping at 50° southward, locally contorted, cut by a fault trending 150°, and showing traces of iridescent copper oxide; the NW–SE fault is marked by ferruginous cherty breccia.

A small roadside pit [6210 2834] near Dinnaton (Figure 10)

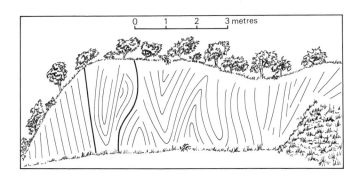

Figure 10 Quarry in Codden Hill Chert at Dinnaton, Swimbridge
Chert beds are disposed in tight and isoclinal folds of about 3-m amplitude, cut by two strike faults. The near-vertical fold axial planes have been bent into gentle flexures

shows cherts striking 100° in tight and isoclinal folds of about 3 m amplitude, cut by two strike faults. The fold axial planes are near vertical, but locally show the effects of a second phase of folding which has bent them into gentle flexures. Exposures in the quarries at Rubble Hills [640 281] show dips of 10°/350° and 12°/215° and also steep and vertical strata trending 100°.

Old quarries just north of East Heddon show cherts dipping at 75°/020° and cut by small dip faults. A NW–SE fault to the east is

visible in an old pit [6580 2883] as a ferruginous cherty breccia. A small quarry [6694 2856] west of Castle Hill (p. 45) shows cherts disposed in an upright near-isoclinal syncline (Figure 11) whose axis is vertical and trends 105° and whose limbs are nearly parallel; the hinge of the fold contains a small fault.

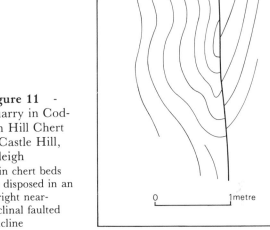

Figure 11 - Quarry in Codden Hill Chert at Castle Hill, Filleigh
Thin chert beds are disposed in an upright near-isoclinal faulted syncline

A small pit [6994 2814] at Aller Cross contains shales and siltstones with cherty bands dipping at 25°/190° and traversed by a bedding-plane fault marked by friable black lustrous shaly gouge. About 1 km to the south, in an old limestone quarry [7000 2716] (p. 45), shales with thin limestones are cut by an E–W vertical fault whose crush zone is up to 0.4 m wide and filled with shaly stony clay gouge. On the southern side of the fault the beds are gently rucked; on the northern side they are contorted and steeply S-dipping (Figure 12). Farther east in the same quarry S-dipping beds contain a bedding-plane fault [7006 2714].　　　EAE, BJW

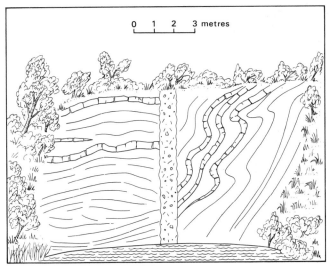

Figure 12 Quarry in Codden Hill Chert at South Aller, South Molton
Shales with thin limestones are cut by an east–west vertical fault marked by a crush zone filled with shaly stony clay gouge

Upper Carboniferous

Sandstones on the south side of Fremington show dips of 45°/350° [5122 3220], 25°/230° [5127 3228] and 10°/200° [5221 3226], suggesting the presence of open or gentle folds.

Laneside exposures [5655 3055] of shales and thin sandstones north of Bishop's Tawton show a fault trending 130°, minor crumpling of beds and traces of southward overfolding. Near the eastern end of the syncline in which lie Venn Quarries, and in the neighbourhood of the fold axis, exposures in the stream valley east of West Combe [6018 3003] show dips of 35°/210° in the northern limb and 20°/010° in the southern limb.

Sandstones and shales in a small stream [6068 2907] east of Hearson are vertical, striking about E–W, or dip steeply to north or south. Similar attitudes prevail in a stream course [6662 2656] and lane [6697 2645] near Bradbury Barton. Sandstone in a quarry [6716 2734] near Meadow Park Farm is nearly horizontal. Exposures [6992 2581] near Kingsland Barton show dips of 75° northward and 80° southward, and similar attitudes prevail along Nadder Lane, South Molton. In Blackrock Quarry [723 251] thickly bedded sandstones dip at 20° northward; they are cut by steep joints trending around NNE and NW, and show slickensides and faulting. The quarry lies between two faults trending ESE. A few similar gentle dips farther east also occur in the vicinity of faults, as at East Mill [7466 2524]. Quarries [7640 2513 to 7680 2516] east of Newton show northerly dips of up to 25° and both northerly and southerly dips of 60° to 70°.   EAE

In that part of the district occupied by Upper Carboniferous strata south of the Nottiston–Codden Hill Lower Carboniferous outcrop, the fold pattern is predominantly of open upright anticlines and synclines. North of Newton Tracey, sandstones and shales of the Crackington Formation dip at 30° to the north [5336 2918], at 70°/355° [5336 2892] and at 70°/180° [5427 2884] near Collabear. Shale overlying Bideford Formation sandstone [5010 2720] 0.5 km S of Horwood dips south at 80°, and 1.5 km to the east [5176 2730] Crackington Formation sandstone dips at 70°/175°. Farther south, Bideford Formation sandstones form two pronounced scarp and dip-slope features [506 264 to 510 265] striking just north of east and suggesting an upright open anticline with a wavelength of about 60 m; these features persist for some 500 m and are abruptly cut off by north-easterly sinistral wrench faults at each end. Bideford Formation sandstones dip at 85°/175° [5218 2578] and at 65° southward [5220 2575] in roadside exposures north of Alverdiscott; east of the village, in the Bude Formation outcrop, black shales dip at 80°/165° at Woodland Cross [5395 2521], and grey shales dip at 70° northward [5451 2506].

East of the River Taw, vertical Crackington Formation sandstones and shales strike E–W at Shilstone Cross [6051 2673], black shales dip at 70° northward [6170 2652] south-east of Cobbaton, and grey silty mudstone dips at 75°/345° at Stowford Cross [6204 2650]. In the Bideford Formation to the south hereabouts, sandstone dips recorded include 75°/175° [6032 2645] near Kewsland, 78° southward [6022 2620] 300 m to the south, 65°/010° [6138 2579] in a road cutting north-east of Hawkridge, and 75°/175° [6153 2612] 400 m to the north-east. At Biddacott grey shales dip at 65°/350° [6281 2588], and 200 m to the east horizontal sandstone crops out in an old quarry [6305 2588]. At Heywood, sandstone dips at 79° southward [6295 2622]; in Chittlehampton, shales and sandstones dip at 45° [6382 2559], 55° [6386 2568] and 50° [6395 2567], all southward, and at Riding Cross sandstone dips at 70°/015° [6475 2577].   BJW

CHAPTER 7

Economic geology

INTRODUCTION

Speculation about the early history of mining and overseas trading in the Ilfracombe–Barnstaple district ranges from the allegation that Egyptian beads were brought to Bampfylde in the Bronze Age to a presumed association between Iron Age earthworks and north Devon iron ores. Dating of the earliest surface workings is impossible. They were almost certainly pre-Roman, though a lump of iron slag from the Dulverton area, now in the British Museum, contains Roman coins. Substantial documentary evidence for medieval mining activity exists, particularly concerning the extraction of silver from the ore deposits around Combe Martin (Figure 13). Transport was always a major problem. Pack-horse trails gave way to lanes and roads, and later a few mines were served by rail, and tramways ran from Florence, Bampfylde and Poltimore mines to South Molton station. Motor vehicles arrived as the industry died, but could not have arrested a decline imposed by world markets and overseas sources.

Building stone and roadstone have been dug from many small pits. Active quarries remain in the Pickwell Down Sandstones, Baggy Sandstones, Pilton Shales and Crackington Formation, and there has been widespread exploitation, on a small scale, of Lower Carboniferous cherts. The large quarries in Pilton Shales between Brayford and Charles (Plate 7) mark the presence within that formation of strong, massive and thickly bedded sandstones. Farther west, hard beds within the shales comprise thin sandstones and calcareous sandstones. True limestones of appreciable thickness occur only in the mid-parts of the Ilfracombe Slates and in the Codden Hill Chert; almost all have been worked in the past (Plate 12).

Sand and gravel are dredged from the Taw estuary and from the sea bed in the immediate vicinity, and are brought by barge to Barnstaple. They comprise variable shelly material for which there is no great demand. The patch of glacial gravel on the outskirts of Barnstaple (p. 49) is too small to constitute a resource, and the pebbly clay and sand west of the town is a mixture of sediment unsuitable for exploitation.

Lake clays south of Bickington extend beyond the present workings for pottery clay, but not sufficiently to sustain extraction on much more than the present small scale.

Peat occurs only on Exmoor. It is thin and remote, too small in amount to consider as a fuel, and low in plant nutrients.

Soils range from rich around the Taw estuary to thin and barren on high Exmoor, and most farming centres on dairying and stock-rearing.

Limited groundwater resources exist in the Hangman Grits, Pickwell Down Sandstones and Baggy Sandstones. Other formations may yield quantities sufficient for single farms. Communal supplies will, however, continue to depend on surface reservoirs and river intakes. EAE

METALLIFEROUS MINING

The Ilfracombe–Barnstaple district includes a large part of the north Devon and west Somerset orefield (Dines, 1956), a metalliferous province probably genetically separate from the tin and copper mineralisation associated with the Cornubian batholith. Silver-bearing lead ores were formerly worked from the Ilfracombe Slates around Combe Martin and also at scattered sites along the outcrop of the Pilton Shales. Deposits of predominantly sideritic iron ores are widespread in the Devonian rocks, and particularly in the Pickwell Down Sandstones. Some of the iron-bearing veins near Heasley Mill carry low-grade primary copper mineralisation, the localised secondary enrichment of which has led to exploitation. Manganese ores have been worked in the district, and occurrences of baryte are known. Traces of gold have been reported from two localities.

The origin of the ores is a matter of debate. Ineson and others (1977) suggested that fluids from the Cornubian batholith might have migrated along the hypothetical Exmoor Thrust (p. 58) during the Variscan orogeny. However, the distances involved and the sparsity of mineralisation in the intervening tract of Carboniferous sediments make this unlikely. No possible local igneous source for the mineralisation is known, although Pattison (1865, p. 226) referred to 'greenstone' near Heasley Mill, and the presence at depth of an igneous body cannot be wholly discounted.

Mineral assemblages throughout the district are indicative of low temperatures of formation. The small size of the orebodies, their commonly concordant nature and their degree of stratigraphical control (Figure 14), all suggest rather restricted movement of ore-forming fluids, in contrast to the formation of the cross-cutting vein systems of Cornwall and south Devon. Together, these characteristics indicate that deposition of the ores was by the localised movement of fluids, of predominantly connate origin, which leached and redistributed low concentrations of metals dispersed within the host sediments (Scrivener and Bennett, 1980). The base metals, particularly lead and zinc, which are virtually restricted to the Lester Slates-and-Sandstones and the Pilton Shales, may have been dispersed originally from exhalative fluids associated with contemporaneous volcanism. The more widely distributed iron ores may have been derived principally from the erosion and redeposition of continental red-bed sediments of the Old Red Sandstone land mass to the north.

Throughout a long history of mining operations the area has returned relatively small tonnages of ore. The deposits are small and patchy and are unlikely to sustain exploitation in the future. The last phase of production was at the end of the 19th century, since which time, apart from a few trials, the orefield has been moribund.

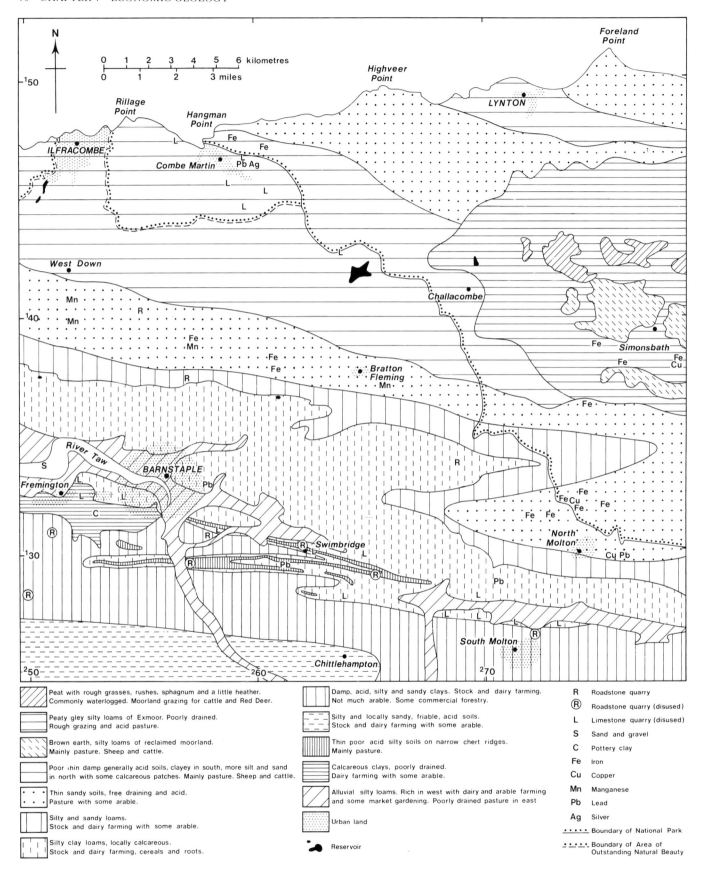

Figure 13 Soils, land use and economic geology

Lead and silver

Records of mining in the area around Combe Martin extend from the late 13th century to the end of the 19th. Early exploitation was under the direct control of the Crown and was largely in search of silver. Tristram Risdon, a 17th century historian whose work was published posthumously in 1811, stated that the Combe Martin mines were founded by Edward I in 1293, when 337 men were moved from Derbyshire to work them. He noted that Combe Martin silver contributed significantly to meeting the costs of the wars against the French in the reign of Edward III. Lysons and Lysons (1822, p. 285) gave details of manning and production for the period 1293–96, but included with Combe Martin the figures for the mines at Bere Alston in the River Tamar valley. Exploitation continued spasmodically thereafter until the last major period of operation in the reign of Elizabeth I, when the mines were worked by various entrepreneurs with the aid of miners imported from the Harz Mountains (Dewey, 1921, p. 56). Since that time, attempts to re-open the workings or to find new deposits have resulted in small returns of ore. Probably the various operations which were financed in the 19th century were little more than trials (Dewey, 1921, p. 57). The main signs of mining activity now visible are at the sites of Combe Martin Mine [5889 4654] and Knap Down Mine [5975 4667].

It seems, from local reports and from notes furnished by Mr H. St L. Cookes, that the earliest workings lay close to the Combe Martin Valley Fault, in the vicinity of the present-day Combe Martin village street. Such evidence as is available suggests that the deposits were patchy, though locally rich.

Examination of mine dumps and coastal outcrops during the recent survey has shown that known lead mineralisation is restricted to the Lester Slates-and-Sandstones. At Lester Point [5750 4761], sulphide ores can be seen in the core of an overturned anticline, forming small irregular masses and lenses streaked out along the cleavage of grey mudstones. The predominant sulphide is galena, which is generally fine-grained and characteristically shows evidence of shearing. It is patchily intergrown with sphalerite and very small masses of chalcopyrite are disseminated throughout. Nearby slates are cut by numerous veins and tension gashes of white quartz, some of which carry aggregates and specks of galena and sphalerite. The quartz is characterised by the presence of small irregular cavities, many of them filled with powdery yellow-brown limonitic material. It appears from comparison with fresh material collected from Knap Down Mine that the mineral originally present was siderite. One cavity yielded a small specimen of a manganese mineral which Mr P. R. Simpson has identified as ranciéite, a calcium-rich variety of psilomelane. Traces of lead mineralisation similar to the occurrence at Lester Point occur at this stratigraphical level along the coastal outcrop of the Lester Slates-and-Sandstones.

The dumps at Knap Down Mine surround an old shaft; they contain grey slates with an irregular cleavage, and fragments of brown-weathering sandstone patchily cemented by siderite. Masses of white quartz bear irregular aggregates of yellowish brown partly weathered siderite, together with galena and sphalerite and minor amounts of chalcopyrite and pyrite. In some large specimens the mineralised assemblage can be seen to form pod-like or lenticular masses which are concordant with the cleavage.

The site of Combe Martin Mine, which was the principal centre of production in the 19th century, is marked by spoil heaps of material similar to that at Knap Down Mine, together with large masses of dark brown siderite carrying stringers of galena. Rottenbury (1974, p. 125) noted that tetrahedrite had been recorded from this mine, but none was found in the recent survey. Mine plans (Nos. R155B and 1222) held in the Mining Record Office at Truro show three orebodies, the principal one, known as Harris's or Combe Martin Lode, being worked from Harris's Shaft [5889 4654] to a depth of 73 m. This structure and the nearby North Lode both strike 080° and dip steeply to the south. South Lode, which strikes 102° and also dips to the south, was intersected by an adit driven northwards towards the workings from its portal [5897 4633] south of Bowhay Lane; only traces of mineralisation were encountered and the dumps have yielded only a few specimens of slate impregnated with pyrite.

Workings at West Challacombe Mine [5858 4734] show no trace of mineralisation and it must be concluded that this was an unsuccessful trial.

The ore minerals of Combe Martin are well represented in the collections of the Geological Museum, London, and of the North Devon Athenaeum, Barnstaple. Specimens of coarsely crystalline galena in these collections characteristically show curved crystal faces, which suggest that the ore has been sheared. Fine-grained galena is commonly intergrown with sphalerite, and study of polished sections has demonstrated that this material also is sheared. Other primary sulphides are pyrite and chalcopyrite; stibnite and millerite have been recorded, but appear to be rare. Secondary species are covellite, malachite and azurite. The tetrahedrite from Combe Martin Mine (Rottenbury, 1974, p. 125) was stated to be argentiferous, and perhaps this mineral was responsible for the high silver value of 5.28 kg per tonne of lead quoted by Dines (1956, p. 757). Dewey (1921, p. 56) referred to filaments of native silver found in galena, presumably in the zone of secondary enrichment.

Such evidence as is available suggests that the Combe Martin lead deposits are not vein structures of Cornubian type. Brecciation is absent, and there is no sign of wallrock alteration or of textures indicating the deposition of ore minerals in open fractures. The orebodies take the form of discontinuous lenses in the cores of near-isoclinal, overturned anticlines. They occur at the stratigraphical level seen in the outcrop at Lester Rock and it is considered that the ores were mobilised from low-grade syngenetic disseminations, and redistributed during the Variscan orogeny. The ores were possibly remobilised and further concentrated by later movements associated with the Combe Martin Valley Fault. A model lead isotope age of 360 ± 30 Ma (Moorbath, 1962) indicates deposition during a late Devonian event. Ages of 294 ± 4 Ma and 308 ± 3 Ma (Ineson and others, 1977, p. 17), based on isotopic work on clays, point to the Variscan orogeny, but do not necessarily relate to deposition of the ore minerals.

A number of trials and small workings for lead exist in the Pilton Shales. The orebodies, in contrast to those at Combe Martin, appear to be crosscutting veins characterised by the presence of minor amounts of arsenic- and antimony-bearing minerals.

The site of a shaft [5780 3300] is all that remains of Pickard's Down Mine. Dines (1956, p. 760) recorded that the dump showed strings of calcite with galena and that some baryte was also present.

An adit, with the portal [6073 2957] to the south-west of Hannaford Bridge, marks the site of East Combe Mine. Though the mine is sited on Lower Carboniferous rocks, debris on the dumps suggests that the workings were partly in Pilton Shales. Galena and sphalerite are present in vein quartz, which cements brecciated shale. Rottenbury (1974, p. 79) noted the presence of minor amounts of arsenopyrite, tetrahedrite and chalcopyrite, the last as inclusions in sphalerite. He also stated that the mine produced some 100 tonnes of argentiferous lead ore in the 19th century.

Old workings [6854 2844] near Shallowford have spoil heaps of shale and veinstone of quartz with galena. Minor amounts of arsenopyrite and an antimony-bearing mineral are also present, the latter stated by Dines (1956, p. 760) to be stibnite or jamesonite.

The largest of the group of lead workings within the Pilton Shales is South Molton Consols, described by Dines under the name of Combe Mine. A lode dipping steeply to the west and striking approximately 145° was worked to a depth of 66 m from a vertical shaft [7006 2853]. The dumps consist of soft grey shale with some calcite-veined limestone, and veinstone of shale breccia cemented with quartz, calcite and siderite. The ore assemblage includes galena and sphalerite with minor amounts of pyrite, chalcopyrite, tetrahedrite and arsenopyrite. The mine was active in the 19th century, and Rottenbury (1974, p. 81) estimated that its production was 437 tonnes of argentiferous lead ore. Badham and others (1979) studied the geochemistry of this site but found no silver in either the galena or the tetrahedrite and concluded that early references to the occurrence of this metal were intended to attract financial support for the mine.

Upcott Mine [7958 2933] was an isolated trial for lead in the Upcott Slates. The dumps consist of slate breccia with quartz, calcite, pyrite, galena and chalcopyrite. There was no known production of ore.

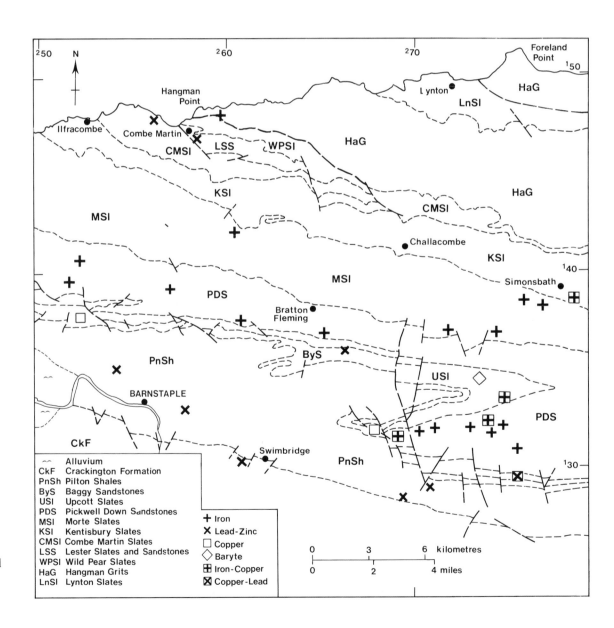

Figure 14 Stratigraphical control of mineralisation

Iron and related mineralisation

Occurrences of iron and associated ores are widespread in the Devonian rocks of the district, and some have been intermittently worked over a long period. Local tradition in north Devon maintains that mining was carried out in Roman times, but the earliest documentary records refer to the production of silver in the North Molton area during the reign of King John and to the production of copper in the same area in 1250 (Rottenbury, 1974, p. 64). A spasmodic output of iron and copper was maintained from the Middle Ages until the period following the industrial revolution, when improved equipment and techniques helped larger-scale mining. A moderate production, mostly of iron ore, was achieved in the 19th century, until competition from more cheaply produced foreign ores brought mining operations to an end. Some recent prospecting has taken place, but has not resulted in productive working.

Information about the nature of the orebodies is scanty, and access to the workings is limited. Most of the deposits are hydrothermal in origin, and were emplaced in fracture systems as localised and irregularly distributed patches of ore. In the northern and central parts of the district the veins are predominantly conformable with the strike and dip, or cleavage dip, of the enclosing sediments. Veins in the southern part of the district, in a belt of country extending eastwards from East Buckland and passing through the copper mining centre of Heasley Mill, cut across the structural grain of their host rocks.

Throughout the district the principal primary ore mineral is siderite, which occurs with quartz as a gangue. Some primary hematite is present, most commonly in the earthy and kidney ore varieties and to a lesser extent in the specular and micaceous forms. Much of the hematite resulted from the oxidation of siderite, as did the mineral goethite. These two oxides formed much of the 19th century production of iron ore; they were called 'brown hematite', and the sideritic material 'spathic ore'. Typically, goethite occurs in brown cellular or massive varieties, but black reniform and botryoidal forms are not uncommon. Yellow-brown limonite and a good deal of clay are commonly present in the weathered ores.

The siderite is of variable manganese content, because rhodochrosite is in solid solution. Segregations of manganese ore, principally pyrolusite, have resulted from alteration of manganese-rich siderite, and these take the form of small veins or localised impregnations of the country rock.

Primary sulphides form a relatively minor part of the assemblage. Pyrite is widely distributed. Chalcopyrite is less common, occurring as patchy low-grade disseminations in siderite. Where such an assemblage has been subjected to oxidation and secondary enrichment, the concentration of copper may have justified commercial exploitation, as around Heasley Mill. Malachite and azurite are the most abundant oxidised species, with chalcocite, bornite and covellite representing the enriched zone. Rottenbury (1974, p. 122) noted that bornite was also found as a primary mineral.

The oldest rocks in the district that contain workable iron deposits are the upper part of the Hangman Grits in the area to the north-east of Combe Martin. An old opencast working [5935 4811] south of Rawn's Rocks is at the westernmost end of a series that extends discontinuously along the strike of the Hangman Grits for about 3 km. Debris near the pit shows strings of goethite in ferruginous sandstone. Nothing is known of the form of the orebodies, and a report by Smyth (1859, p. 106) that they consisted of clay ironstone nodules has not been substantiated. Dines (1956, p. 758) quoted a total production of approximately 11 500 tonnes of brown hematite and siliceous siderite from North Devon Iron Mine at various times in the 19th century.

The Ilfracombe Slates show only a few scattered trials for iron ore, and production figures are unknown. Weathered ferruginous limestones in the Combe Martin Slates were formerly worked for the mineral pigment umber, and De la Beche (1839, p. 646) reported that a large quantity of ochre was raised in the parish of East Down. The workings of Wheal Eliza [7847 3811] are in the Kentisbury Slates, close to the boundary with the Morte Slates. Two parallel E–W lodes which dip at 75° southwards were explored to a depth of 91 m. The mine is not typical of the district in that copper was the principal metal sought. Ore on the dumps consists of granular siderite with disseminated chalcopyrite and pyrite. Some massive pyrite is also present. Secondary chalcocite, malachite and pyrolusite are found in association with goethite and hematite. Early reports in the Mining Journal (1848, p. 120; 1854, p. 206) that baryte was present have not been confirmed. Production figures for Wheal Eliza are unknown, but Orwin (1929, p. 117) reported that the run-of-mine ore contained 11 per cent copper and 60 per cent iron.

Iron ore deposits occur in the eastern part of the outcrop of the Morte Slates. The largest group of workings extends along the strike of the slates between Cornham Ford [7490 3864] and a locality some 400 m ENE of Blue Gate [7581 3767]. Deerpark Mine is the western part of this complex and the workings to the east of Blue Gate are known as Exmoor Mine. The main source of ore at Deerpark Mine was Roman Lode, which was worked from the 46-m-deep Roman Shaft [7495 3828]; it strikes 110° and dips steeply to the south, conformably with the bedding of the enclosing sediments. A section of the lode seen in a pillar and recorded by Dines (1956, p. 764) was 2.29 m wide and consisted of siliceous goethite and hematite with bands of chert and granular quartz. Dines also stated that some 650 tonnes of ore were produced from Roman Lode during a period of working that ended in 1856, but the nature of the openworks suggests that the mine is of considerable antiquity.

Exmoor Mine worked two structures: Double Lode, possibly an extension of Roman Lode; and Rogers Lode, which trends 090° and meets Double Lode at 200 m NE of Blue Gate. An attempt was made to work this mine between 1910 and 1913, when 1727 tonnes of 56 per cent iron ore were raised. In an unpublished report dated 1940, Dr A. W. Groves quoted the following partial analyses of ores:

	Deerpark Mine %	Exmoor Mine %
Fe	64.80	56.13
Mn	0.06	3.71
S	0.026	0.021
P	0.029	0.045
H_2O	3.85	8.35

In Hangley Cleave an adit [7441 3661], driven south to intersect a lode which strikes 063° and dips to the north-west, produced 1220 tonnes of ore in 1853 (Dines, 1956, p. 764).

Spoil from the adit consists of quartz and siderite. Other workings in the Morte Slates are either trials or ancient sites, such as that at Mole's Chamber [7205 3957], for which there are no records of production.

Within the Pickwell Down Sandstones, small deposits of iron and manganese in the western and central part of the outcrop give place to larger scale mineralisation farther east. Some degree of structural control is suggested by the concentration of the Heasley Mill copper–iron deposits on the northern limb of the Bray Valley Anticline.

Fullabrook Mine [5176 3990] was a trial for manganese on two parallel lodes which strike 015° and dip steeply to the east. The ore consists of sandstone breccia cemented by pyrolusite and wad. A similar occurrence was tried at Snowball Hill Mine [5179 4091], where dumps show ramifying veinlets of manganese ore in red- and purple-stained sandstone.

Viveham and Shirwell mines were trials for iron ore which may have produced some mineral pigment. A line of shafts and open works at Viveham Mine [5677 3889 to 5718 3884] affords little evidence of the nature of the deposits, but the veins appear to strike NW–SE. The same trend is shown at Shirwell Mine, where workings followed two veins and 41 tonnes of iron ore were produced in 1873 (Dines, 1956, p. 759). The northern working [6069 3833] is reported by the same author to have been driven on thin bands of ochre but the southern vein, explored by a shaft [6069 3774], appears to have been barren.

Haxton or Bratton Fleming Mine [6557 3711] was a trial for iron and manganese. The spoil heaps show pink sandstone with veinlets and botryoidal masses of manganese ore. A number of trials for iron, manganese and copper were made in the area to the north-east of East Buckland, extending eastwards from Mill Wood [681 316] across Wrick Down. A lode explored by adits and a shaft [6808 3165] is stated by Rottenbury (1974, p. 60) to contain siderite with spots of bornite. It strikes about 085° and lies within the sett known as Wheal Charles. Pits [6875 3184] on Wrick Down, which are known as Orestone Mine (Rottenbury, 1974, p. 60), were said to have been worked for iron and manganese, but there are no details of production.

At a trial known as Poltimore Mine, an adit [7010 3205] some 200 m N of Walscott Farm was driven to intersect two lodes trending somewhat south of east. The ore consists of siderite with disseminated specks of chalcopyrite.

Stowford Mine is marked by a line of shafts, adits and openworks [7111 3206 to 7152 3185] which, according to Rottenbury (1974, p. 150), exploited three lodes. Much of the ore appears to be oxidised siderite, but an unusual feature is the presence of micaceous hematite which may be primary. Some 5000 tonnes of ore, mostly brown hematite, were produced in the 19th century. The ancient workings of Barton Pits extend over almost 1 km [7211 3207 to 7290 3198] and consist of three groups of openworks arranged en échelon. Ore specimens from this locality are mostly of hematite, both massive and specular, with some goethite. Rottenbury (1974, p. 63) recorded that in the 19th century a small amount of manganese was raised both here and at Crowbarn Mine [7385 3182], which is on strike with the eastern workings of Barton Pits and about 1 km to the east. The spoil heaps at Crowbarn Mine show massive and specular hematite intergrown with comby and vughy quartz.

Bampfylde Mine, also known as Poltimore Mine, is the largest working in the district, having yielded an estimated 15 750 tonnes of 15 per cent copper ore since 1696, and a considerable tonnage of iron ore worked in early times from the gossan (Rottenbury, 1974, p. 71). The workings span the River Mole valley some 600 m N of Heasley Mill, with the main area of production on the western side. Poltimore Lode, the most productive of the three structures worked, trends roughly E–W and dips to the south at about 80°. North Lode and Bampfylde Lode are parallel with Poltimore Lode in strike and dip steeply north and south respectively. A shaft [7538 3273] serviced the western part of Poltimore Lode to a depth of 205 m; others, sunk on the eastern side of the valley, reached a maximum depth of 110 m. The ores occur in well-defined fissure veins which are cut by numerous cross-courses. Quartz and specular or micaceous hematite are the predominant ore minerals with siderite at depth. Copper is represented by malachite and azurite in the oxidised zone and by chalcocite, bornite and chalcopyrite beneath. The copper ores at Bampfylde Mine were reputed to be argentiferous, and there are reports quoted by Pattison (1865, p. 225) of gold in the gossan. Rottenbury (1974, p. 167) confirmed the presence of low concentrations of gold in North Lode, the average value being around 17 ppm.

The largest producer of iron within the area of outcrop of the Pickwell Down Sandstones was Florence Mine, where 39 000 tonnes of sideritic and hematitic ores were raised between 1873 and 1885. The two main structures, North Lode and South Lode, trend 115° across Radworthy Down Cleave some 330 m apart. Both veins are nearly vertical, and South Lode was serviced by a shaft [7518 3200]. A partial analysis of hematite ore from this mine was quoted by Dr A. W. Groves (op. cit.) as follows: iron, 58.83 per cent; manganese, trace; sulphur, 0.077 per cent; phosphorus, 0.009 per cent.

Deposits in the Upcott Slates are less well developed than in the underlying Pickwell Down Sandstones but show a greater range of metals. The occurrence of lead with copper and iron at Upcott Mine has already been noted, and the mineralogically diverse orebodies of Bampfylde Mine occur, in part, in this formation. Britannia Mine [7455 3359] was a trial on an E–W orebody which carried iron, some copper and traces of gold. Pattison (1865, p. 226) stated that the gold was found in association with hematite and that greenstone was present in the lowest rocks. These observations have not been confirmed. Rottenbury (1974, p. 124) recorded baryte, chalcopyrite and tetrahedrite on the dumps, and also noted a more important occurrence of baryte some 400 m N of Bentwichen. At the latter locality a vein extends 750 m eastward from Span Bottom [7278 3447]; the eastern part of the lode is offset to the south by a NW–SE fault. Trenches have shown that the lode is nearly vertical and ranges up to 2 m in width. The principal minerals are baryte and quartz, with minor iron oxides and scattered pockets of bornite crystals.

Mineralisation is very sparse in the Baggy Sandstones. An old trial for copper, known as Beer Charter Mine, is marked by an adit [5319 3761], and a small working for the same metal near East Buckland by a shaft [6700 3186]. Lead was said by Rottenbury (1974, p. 59) to have been the object of a trial [6657 3588] on Mockham Down.   RCS

BUILDING STONE AND ROAD STONE

Little Silver Quarry [549 403], owned by Kingston Minerals Ltd, is cut in fine- to medium-grained sandstones of the Pickwell Down Sandstones. Some 105 m of strata are exposed, dipping at moderate angles to slightly west of south, and the workings progress westwards along the strike. About 75 000 tonnes of stone are quarried annually; products are road aggregate and a little building stone.

Plaistow Quarry [568 373] exhibits 60 m of fine- to medium-grained sandstones with some shales and siltstones (Figure 3) of the Baggy Sandstones, dipping at about 50° southwards and worked eastwards. The main bay shows a 46-m-high face, and friable sandstones with interbedded shales at its southern end are the stratigraphically highest beds exposed; two faults occur at its northern end, and sandstones and shales partly obscured by scree separate it from a currently disused bay to the north. Road aggregate is the main product. Archibald Nott and Sons, the owners, quote a variable annual production of up to 20 000 tonnes or more, including up to 300 tonnes of building stone.

Several quarries [e.g. 687 337], worked by the same company in the valley of the River Bray between Brayford and Charles (Plate 7), are in thick hard strong sandstones within the Pilton Shales. In contrast to Plaistow and Little Silver quarries, where fairly uniformly dipping strata occupy the whole workings, the beds in the Bray valley quarries are disposed in E–W trending folds of amplitudes commonly less than the heights of the faces. Working proceeds westwards along the strike, but problems occur locally where fold plunges bring in increasing thicknesses of thinly bedded sandstones and shaly strata. The main product is road aggregate. Some 75 000 to 80 000 tonnes of stone are extracted annually; about 20 000 tonnes are sold ready-coated with bitumen, and output also includes rather less than 200 tonnes of building stone.

The extensive Venn Quarries [581 306] of E.C.C. Quarries Ltd show a long strike section in fine-grained sandstones with interbedded shales near the base of the Crackington Formation (Plate 13). Some sandstone beds are up to 1.5 m thick but most are much thinner, and the large amount of interbedded shales leads to the accumulation of great quantities of waste. The sandstones have a high polished stone value and are used mainly for road surfacing aggregate. Production is about 100 000 tonnes per year.

Plate 13 Venn Quarries
Bedding planes of Crackington Formation sandstones. The sandstone is worked for roadstone, the interbedded shales discarded. (A 11839)

Disused quarries and pits mark the many places where stone has been dug for local use in roads and buildings, poor quality local material commonly having been used in preference to better stone from some distance away. Only the larger pits are recorded here.

Lawns Quarry [656 437], on the northern slopes of Rowley Down and now an overgrown rubbish tip, was a narrow E–W cut 120 m long and presumably followed a thin sandstone within the Combe Martin Slates.

Sandstones within the Kentisbury Slates have been worked in Greatgate Quarry [6130 4250], a pit [6223 4425] north of Kentisbury, and numerous small workings on the high ground of Kentisbury Down, Rowley Down, Challacombe Common, North Regis Common, Broad Mead, Goat Hill, Dure Down, Great Ashcombe and Little Ashcombe.

A 6-m face of Morte Slates in an old quarry [5000 4348] at Trimstone affords an example of poor-quality building stone being dug on site to avoid transport costs. There are several shallow old quarries in Morte Slates on Bratton Down, formerly sources of material for dry-stone walling, amongst them one with a 3-m face [6625 3917] at Knightacott Cross. Similar quarries on Wallover Down include a 6-m exposure of slate [6895 3924]. The disused Castle Quarries [705 388] on Castle Common display the weathered remains of slate faces up to 3 m high, again used locally. Scattered old quarries are found on the high ground south of Simonsbath, and a series of them borders the South Molton to Simonsbath road, displaying faces 8.5 m [7606 3806], 4.25 m [7608 3828], 15 m [761 385] and 10 m [7628 3869] in height.

Several small pits in fine-grained sandstones of the Pickwell Down Sandstones occur in the vicinity of Buttercombe Wood; two of the largest show 5.5 m [5042 4199] and 21 m [5069 4185] of strata. This formation has been extensively pitted for local use [5310 3932; 5991 3784]. Sandstones in the upper part of the formation have been dug between Loxhore Cott and Lower Loxhore [6102 3813; 6094 3778]. The impersistent but massive tuff which marks the base of the formation has been much used [5502 4086; 5970 4008; 5986 4005; 6324 3894].

Berryhill Quarry [669 372] and a small quarry [6675 3720] along the strike to the west worked hard sandstones, probably for building stone. To the east, in the Bray valley, other smaller quarries worked Pickwell Down Sandstones for local use around Wort Wood. Several small quarries provided local building material around Stowford Bridge; of these, two [712 318 and 713 316], each display a 3.5-m face of sandstone. Farther east, only very small quarries are found on the high ground north of Kerswell Farm.

The Baggy Sandstones contain numerous impersistent fine- and medium-grained sandstones, mainly up to a few metres thick, within a sequence of interbedded sandstones, siltstones and shales. Thus small pits are commonly in sandstone [5055 3778; 5360 3756; 5364 3743; 5500 3736; 5668 3747; 6162 3661], whereas the larger Plaistow Quarry [568 373] cuts into a good deal of softer thinly bedded and argillaceous strata.

A large old quarry within the prehistoric camp at Mockham Down [666 358] is water-filled, but a 16-m face of sandstone can be seen. Small old quarries are found sporadically on the Baggy Sandstones outcrop around the Bray valley folds, and farther east, south of North Molton, some larger quarries occur. These include one at Nadrid Copse [704 295], and a large quarry in Portfolken Wood [7276 2918] that reveals a 12.3-m face in hard sandstones with shaly ribs, which was worked within the last 20 years.

Sandstones up to 2.5 m thick and locally calcareous have been dug in a few places in the western crop of the Pilton Shales [5657 3602; 5660 3592; 5744 3611; 5598 3508; 5688 3532; 5783 3485; 5948 3473; 5943 3448; 6079 3581]. In one small quarry [5620 3512] north of Barnstaple the hard beds are of calcareous sandstone and sandy limestone; although probably dug for roadstone they could possibly have been burnt for lime.

East of the main concentration of sandstone quarries within the Pilton Shales of the Bray valley, there is a scattering of small quarries in splintery grey sandstones between North Molton and South Molton, particularly in the valley west of Burcombe [notably 7236 2837 and 7244 2870].

Lower Carboniferous cherts have been quarried sporadically throughout the length of their outcrop. They are less hard than is commonly the case with cherts, and too thinly bedded and closely jointed to yield large blocks, but their upstanding topographic ridges attest strong resistance to erosion, and the rocks have been much used in the making of roads and tracks. No active quarrying takes place at the present time.

Cherts have been dug from several pits on the ridge from Rumsam to Landkey; one of them [5725 3134] shows over 6 m of strata. The same belt of rocks has been worked near Swimbridge exposing 8 m [6155 3028], 3 m [6166 3023] and 20 m [6201 3013] of strata, mainly fairly thinly bedded cherts. A few quarries mark the extraction of cherts west of the River Taw, for example, Templeton Quarry [5432 2972] and Park Gate Quarry [5552 2970], but the biggest chert quarry [569 297] in the district lies just east of the river at the western foot of Codden Hill and shows some 20 m of beds. Also on Codden Hill are Overton Quarry [5735 2944], with 10 m of chert exposed, a quarry south of Codden Beacon [5829 2942] with 12.2 m of chert, and Tower Hamlets Quarry [5965 2935], now inaccessible. Ridges south of Swimbridge have been pitted locally [6030 2984; 6213 2952; 6240 2950].

East of Swimbridge cherts have been dug in a pit [631 296] near Kerscott and in others [6474 2940; 6498 2938] near Heddon. A larger quarry [653 290] in banded cherts lies just east of Heddon, and small pits occur sporadically on the ridge to the east as far as Deer Park Gate [6770 2856].

Many small disused quarries in sandstones are scattered across the Upper Carboniferous outcrop. Sandstones in beds up to 1 m thick have been worked [5077 3115] in Bickleton Wood. Old Quarries [5597 3080; 5640 3110; 5639 3103] alongside the River Taw south of Barnstaple show up to 12 m of sandstones in beds up to 1 m thick with interbedded shales.

In a pit on the western edge of the district [4978 2844], 7 m of thickly bedded sandstones lie within a succession of thinly bedded sandstones and shales. In the south-western part of the district a few small sandstone pits occur, as south of Horwood [5011 2722; 5060 2643]; a 5-m face in blocky brown sandstone occurs in a pit [5311 2830] east of Prospect Place.

East of the Taw valley, several quarries have been worked at Chittlehampton [6378 2568; 6380 2524]. Farther east 2.5 m of massive sandstone have been worked [6716 2734] south-east of Filleigh, and a larger quarry [719 266] north-east of South Molton exposes fine-grained sandstone in beds up to 0.5 m thick, with some siltstone. Other old quarries showing massive fine- and medium-grained feldspathic sandstones occur locally [6701 2628; 6795 2654; 7209 2628; 723 251; 7533 2539; 7640 2512 to 7679 2515].   EAE, BJW

LIMESTONE

Substantial amounts of limestone have been dug from the Ilfracombe Slates and the Codden Hill Chert. The main limestones of the Combe Martin Slates are persistent, the thickest two comprising up to 10 m of beds. Otherwise, limestones within the Ilfracombe Slates are uncommon and lenticular. The worked limestones of the Codden Hill Chert are invariably lenticular; they have been dug to their economic limits, laterally to the edges of the lens and vertically either to similar margins or, more usually, to a depth determined by inflow of water. There is some evidence to suggest that in the eastern part of the crop a mixed succession of shales, siltstones and sandstones, locally calcareous and cherty, with small limestone lenses, has been quarried for lime. Thin limestones are plentiful within the Pilton Shales but have proved workable only in the few places where they are aggregated into thick lenses.

Disused limestone workings in the Combe Martin Slates near Combe Martin, most containing the remains of lime kilns, include the following: Napps Quarry [565 476], quarry [579 469], Harris's Quarry [580 466], quarry [583 466], Park Quarry [584 462], Rectory Quarries [584 456], Eastacott Quarries [586 456], Locks and Tracey Down Quarry [591 456] and Berry's Quarry [594 456]. Henstridge Quarry [591 448] lies near the top of the Combe Martin Slates. It is fairly large but much overgrown with vegetation. An old track down to it from the nearby road is so carefully routed and graded as to suggest long-continued traffic of horse-drawn carts. The rocks are slates with thin sandstones and thin bands and lenticles of grey limestone. Beds of this nature would not have been dug from such a place for roadstone or building stone, and it is presumed that the quarry was operated as a source of lime, although there is no trace of such thick limestones as occur farther north. Certain remains deep in the nearby valley may have been limekilns.

Comer's Ground Quarry [6405 4243], near the top of the Kentisbury Slates, marks the working out of lenticular limestones which were burnt in a kiln on the site. No other occurrence of limestone is known in this formation.

A large overgrown and flooded quarry [6477 3000], in Pilton Shales 2.75 km E of Swimbridge, appears to have been opened for limestone; lenses of sandy limestone and grey crystalline limestone lie within shales and were dug and burnt on the site. A smaller old limestone quarry [6445 2972] to the south-west, alongside the old railway track, is probably also in Pilton Shales.

Thin limestones are common in the Lower Carboniferous rocks. They occur within the shaly sequences which both underlie and overlie the cherts and which, in the west, constitute the whole Lower Carboniferous succession between the Pilton Shales and the Crackington Formation. Local thicker lenticular limestones occur sporadically, and every one of them has been quarried in the past. No limestone has been worked later than the first quarter of this century.

About 4 m of thickly bedded limestones have been worked [519 332] near Penhill, and several quarries [519 332] near Penhill, and several quarries [5191 3251 to 5226 3246] were once active near Muddlebridge. A flooded quarry [5425 3227] occurs in the western outskirts of Barnstaple, and a larger one [554 315] at Lake, just south of the town. Spoil at Lake has been tipped to form 'Donkey Hill', which bears traces of a spiral track to the top. Donkeys were used to carry both limestone and waste.

Old quarries [581 309 to 588 308] at Venn, just north of the active sandstones quarries (p. 75), were opened in limestone lenses up to 20 m thick. They have been largely filled but some exposures remain, as at the western end.

Marsh's Quarries delineate a narrow lenticular group of thickly bedded limestones [606 304 to 614 301] and are currently being filled. Bestridge Quarry [625 299] was a similar working at the same stratigraphical level, and lenticular limestones have also been exploited at Rubble Hills [640 281].

East of the River Bray, cherts are not separately mappable and small quarries within the northern crop of the Codden Hill Chert show shales, siltstones and sandstones, locally siliceous and cherty. Lenses of limestone range up to 0.7 m thick near North Aller [696 282], and it appears that these beds were worked primarily for lime. The southern crop contains more and thicker lenticular beds of limestone which have been extensively quarried [6745 2759 to 7019 2707; 7091 2708 to 7123 2708; 7218 2699 to 7298 2696].

SAND AND GRAVEL

The only sands and gravels of economic interest in the district are the mixed shelly deposits of the estuary floor. They are dug at low water and brought by barge to Barnstaple on the rising tide. Material from farther downstream, and occasionally some dredged from off-shore, is also brought to the town.

CLAY

Glacial lake clays (p. 49) are worked at Higher Gorse Claypits, south of Bickington. They are used at the Litchdon Pottery of C. H. Brannam Ltd, in Barnstaple, to produce decorative domestic ware, but have also been utilised for bricks, drain-pipes, flowerpots and cloam ovens.

Evidence from present and disused pits, and from a number of shallow boreholes, suggests that stone-free clays extend throughout the area of Claypit Coverts, slightly beyond that area to the north, south and west, and as a tapered easterly extension for almost 500 m. They lie beneath 3 to 5 m of head and boulder clay. The greatest thickness, of about 6 m, occurs around the present workings and for some 300 m northwards. Reserves may total about 1 800 000 m^3.

Boulder clays with scattered, generally small, stones extend over a larger area from Fremington to Roundswell and constitute a reserve of brick clay of variable quality. Thin smooth clays occur locally, and some were encountered at a depth of 12.2 m in a borehole [5415 3125] at Roundswell.

Two brickfields were in operation in Barnstaple in the 19th century, one at Pottington, west of the town and near the Braunton road, and the other at a site now covered by Summerland Street and adjoining buildings [562 330]. The latter pit showed at least 4.3 m of clay (p. 51). It is uncertain whether the clays at Pottington occurred in the 1st Terrace or in the modern alluvium, but those of Summerland Street lay within the 1st Terrace.

PEAT

Peat in mappable thicknesses is confined almost entirely to high Exmoor. The largest spread, over The Chains and Hoaroak Hill, extends to almost 2.5 km^2 and is up to 2.6 m thick. A more ramifying cap covering about 1.6 km^2 on Exe Plain and Lanacombe is rarely more than 1 m thick, and smaller patches on Winaway (0.25 km^2) and Trout Hill (0.7 km^2) are generally 0.5 m or so thick.

The peat forms blanket bog and is an acid type (around pH 3.5) composed mainly of sphagnum moss, deer-grass, cotton-grass and purple moor-grass (Curtis, 1971). Traces of old diggings remain on Hoaroak Hill, Exe Plain and Lanacombe as shallow depressions, damp or filled with water. This early working of the peat was for fuel. The small amounts present, remote situation, and usefulness of the peat as a regulator of run-off, combine to suggest that no future exploitation is practicable. EAE

COAL ('CULM') AND MINERAL BLACK

Traces of old pits, adits and shafts found near the southern border of the district mark workings in carbonaceous material which extends from the coast at Greencliff, in the adjacent Bideford district, eastwards to Hawkridge Wood on the eastern flank of the Taw valley. It is likely that output was very small, but one shaft in Hawkridge Wood [6054 2514] was reputedly sunk to a depth of about 31 m. The history of the industry is summarised in Edmonds and others (1979). BJW

SOILS

Soils and the pattern of farming which they support are determined by two main influences: the various rock formation, which mostly trend east or east-south-east, and the general increase in altitude of the land surface northwards and north-eastwards to Exmoor. Local modifications reflect local physiography, such as deep-cut valleys or steep slopes, which in turn effect different changes according to their alignment and aspect.

Much of the arable farming is confined to the slightly calcareous silty clay loams, locally sandy, overlying the Pilton Shales, although dairy farming is the dominant activity, as it is on most of the lower ground. Carboniferous rocks younger than the Pilton Shales weather to mainly clays and silty clays. Some sandy soils on low well-drained ridges of Upper Carboniferous sandstone support cereal crops locally, but most of this area is devoted to dairying and raising stock, as are the boulder clay soils south of the estuary.

Farther north the interbedded sandstones and shales of the Baggy Sandstones show a few arable fields on fine sandy loams. Generally, however, the soils on the older Devonian formations are thin and acid. They support some dairy herds in their southern and western parts, but sheep and beef-cattle become dominant towards Exmoor.

The Pickwell Down Sandstones are covered mainly by free-draining acid brown-earth soils comprising silty loam on stony silty loam; such stony soils do not retain water, and they are largely given over to pasture. Similar soils occur on the silty and sandy parts of the Morte Slates and Ilfracombe Slates; drainage is poorer and the land use is as pasture. However, the main rock of these two formations is slate, which commonly yields peaty gleyed podzols in which a thin peaty surface overlies leached silty loam; iron pan is present at shallow depth in many places. In some localities these soils are undisturbed and bear rough moorland grasses; in others, as near Simonsbath and around some of the moorland farms, they have been reclaimed by ploughing to break up the iron pan and they carry acid pasture. Both types afford grazing for sheep and cattle. The Hangman Grits of the high moor weather to free-draining podzols in which a thin peaty layer overlies sandy loam. Some reclamation has provided cultivated acid pasture, but in the main the ground is covered by coarse moorland grasses with some heather.

The blanket bog peat of the high moor affords only very rough grazing, of coarse grasses and a little heather, for a few Galloway cattle and the native red deer.

Steep valley sides on all formations are commonly wooded, generally with deciduous plantations, but locally and increasingly with commercial plantings of conifers. Aspect is particularly important on the high ground, where only the south-facing slopes offer much scope for settlement and reclamation.

WATER SUPPLY

No major aquifer exists beneath the district, and no large supplies of groundwater are available. However, small quantities are drawn from wells and boreholes, mainly by individual users. Slates and shales of the Ilfracombe Slates, Morte Slates, Upcott Slates, Pilton Shales and much of the Carboniferous succession are of low permeability. Fractures produced by tectonic movements store some groundwater, and a chance intersection by a borehole may result in moderate initial yields and rapid replenishment. Unfortunately when the reserves in interconnected fissures are pumped out, further inflow is dependent on intergranular movement of water and yields drop sharply. Supplies of up to 3180 litres per hour (l/h) have been obtained from these rocks from depths to 60 m, although around 1135 to 1800 l/h is more typical of the successful sinkings.

Sandstones of the Hangman Grits, Pickwell Down Sandstones and Baggy Sandstones contain some water, although they are commonly strongly cemented. Fissures probably constitute most of the available storage, and although the competent nature of the rocks has resulted in more resistance to tectonic movement than in the case of argillaceous strata, the fewer fractures formed are more likely to have remained open. Yields from five boreholes in the Hangman Grits, with depths to 35 m, have ranged from 1135 to 3180 l/h. Few figures are available for the other sandy formations. A 64-m borehole [497 376] near Boode, on the western edge of the district, sunk for a public supply, probably passed through basal Pilton Shales into Baggy Sandstones. It yielded 14 500 l/h in an initial pumping test, and subsequently 227 000 litres per day (1/d) for periods of about three weeks.

Two boreholes [503 348] at the old R.A.F. station at Chivenor proved 15 m of alluvial clays, sands and gravels on Pilton Shales. They are recorded as having been test-pumped at 12 270 and 29 530 l/h. Inflow was from the alluvium, and consisted of water in hydrological continuity with that of the River Taw but separated from it by a width of sediment sufficient to permit natural bacteriological purification.

A small reservoir [505 376] alongside Buttercombe Lane, on Pilton Shales in the west of the district, was constructed in the 19th century to supply Braunton by gravity feed. Ilfracombe was served from about 1800 onwards by two reservoirs at Slade, on Morte Slates, with a combined capacity of 245 million litres. In 1940 additional storage was provided in the upper reaches of the River Bray upstream of Challacombe by the construction of a dam 107 m long and 14 m high. The bedrock of Kentisbury Slates was found to be much fissured and to require extensive grouting. About 68 million litres are impounded. A larger dam at Wistlandpound, 162 m long and 28 m high, is founded on Morte Slates; the reservoir extends on to Kentisbury Slates and stores 1550 million litres of water. Water is supplied from Wistlandpound to that area of north Devon lying north of Chulmleigh and Witheridge, west of the River Exe, east of the River Taw and south and west of Exmoor. Barnstaple receives about 1.3 ml/d, from Wistlandpound Reservoir, but draws its main supply, about 4.5 ml/d, from Loxhore Pond, a small reservoir at Loxhore Bridge on the River Yeo north of Chelfham.

FUTURE PROSPECTS

There is little prospect of a resurgence of mining in the district. Exploration in the 1970s in the area north of North Molton has led to nothing. There are no iron ores in the district worth exploiting in the foreseeable future; Britain possesses large resources of more worth than those of north Devon. Of the rarer metals it is probable that some copper lodes associated with known old workings have not been fully exploited, but it is unlikely that ores of this sort are worth seeking in the district at present. Only the discovery of some major parent source, perhaps of disseminated ('porphyry') copper, would hold out the prospect of a new, and different, phase of metalliferous mining.

Resources of 'gritstone', however, are huge. They lie mainly within the Pickwell Down Sandstones and the arenaceous strata of the eastern part of the Pilton Shales. Sandstones of the Baggy Sandstones are commonly impersistent, and those of the Upper Carboniferous generally interbedded with shales. Nevertheless all these formations contain large resources, which could be of use in local construction, mainly as road surface aggregate, even though transport beyond the district may be uneconomic. Lower Carboniferous cherts are likewise available, although commonly softer than such rocks generally are.

Probably the lenticular limestones of the upper Ilfracombe Slates and the Lower Carboniferous have been more or less worked out. The more persistent limestones within the Combe Martin Slates, none of which is currently quarried, afford some resources.

Sand and gravel in workable amounts occur only offshore. Glacial lake clays south of Fremington and Bickington will continue to support the small pottery industry, and adjacent slightly stony boulder clay is probably locally suitable for brick-making, although production of bricks from the much larger clay reserves of Meeth and Petrockstow was not sustained. Peat on Exmoor is too thin, patchy and remote to tempt potential diggers.

No large resources of groundwater are present beneath the district, and any major scheme to augment supplies will necessarily rely on surface storage or river intakes. Old reservoirs have been supplemented by dams at Meldon on northern Dartmoor and Wimbleball on the southern edge of Exmoor, and other possible sites exist. Storage of large quantities of fresh water in or near the district would, however, be possible only in the valleys of the rivers Taw and Torridge or in their estuary.   EAE

CHAPTER 8

Geophysical investigations

REGIONAL GRAVITY AND MAGNETIC SURVEYS

Marine gravity data for the Bristol Channel were collected jointly by the University of Swansea and the Institute of Geological Sciences in 1971 (Brooks and Thompson, 1973); north–south traverse lines were 5 km apart on average, and the mean discrepancy at the intersections with tie lines was 1.1 mGal. On land, regional gravity measurements were carried out by Dr A. J. Burley in 1970, at a station density of a little under one per square kilometre; their accuracy is better than 0.1 mGal. All observed gravity values have been adjusted to the 1973 National Gravity Reference Network (NGRN 73), and the Bouguer anomalies shown in Figure 15

have been derived using the 1967 International Gravity Formula and rock densities of 2.67 t/m³ (marine surveys) and 2.70 t/m³ (land surveys). Terrain corrections were applied to land data.

Figure 16 is a total magnetic field map, expressed as departures from a linear reference field having a value of 47 033 nT (nanotesla ≡ gamma) at the origin of the British National Grid, increasing uniformly by 2.1728 nT/km in the direction of grid north and 0.259 nT/km in the direction of grid west. Two separate airborne surveys were flown for the Survey by Hunting Surveys Ltd: in 1958 over land areas along north–south lines 0.4 km apart at a terrain clearance of 152 ± 30 m, and in 1961 over the sea along

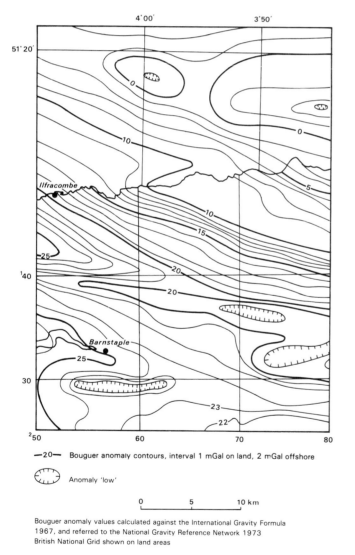

—20— Bouguer anomaly contours, interval 1 mGal on land, 2 mGal offshore

Anomaly 'low'

0   5   10 km

Bouguer anomaly values calculated against the International Gravity Formula 1967, and referred to the National Gravity Reference Network 1973
British National Grid shown on land areas

Figure 15 Bouguer anomaly map of the Ilfracombe and Barnstaple districts

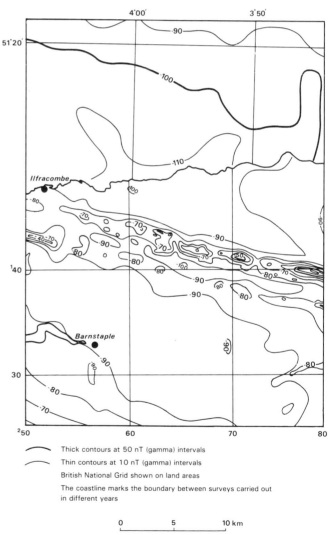

Thick contours at 50 nT (gamma) intervals

Thin contours at 10 nT (gamma) intervals

British National Grid shown on land areas

The coastline marks the boundary between surveys carried out in different years

0   5   10 km

Figure 16 Aeromagnetic map of the Ilfracombe and Barnstaple districts

north–south lines 2 km apart at a height of 305 ± 46 m. The mis-match between the contours at the coastline probably results from the difference in flying height.

GEOPHYSICAL ANOMALIES

Bouguer anomaly values decrease northwards across Exmoor and the southern Bristol Channel, reaching a minimum along an ESE–WNW line marking the axis of the Bristol Channel Syncline (Lloyd and others, 1973). Seismic refraction studies (Brooks and James, 1975) have shown that the Mesozoic succession preserved in the Bristol Channel Syncline is probably 2000 to 2400 m thick, but these low density rocks can explain only a part of the negative gravity anomaly. Even if a steep regional gradient is arbitrarily assumed, the shape of the resulting residual anomaly shows clearly that a wider or deeper source is responsible (Brooks and Thompson, 1973, fig. 6). Several possible interpretations have been advanced:

1 A deep crustal explanation, such as a slight depression in the Moho or a slight progressive change in mean crustal density, is unsatisfactory because consideration of the anomaly amplitude and maximum gradient suggest a source at not more than 11.4 km depth (Bott and others, 1958) and because of the markedly different regional gradients in surrounding areas (Brooks and Thompson, 1973).

2 An unexposed granite might account for the size of the anomaly but not its shape: the steady gradients and the shape of the contours are not characteristic of granite masses (Bott and others, 1958).

3 A southward-thinning wedge of low density Lower Devonian or pre-Devonian rocks stratigraphically underlying the oldest exposed Devonian strata (Lynton Slates) would need to be from 2 to 6 km thick (depending on the choice of density contrast and regional field) to explain the observed anomaly (Bott and others, 1958; Bott and Scott, 1966; Brooks and Thompson, 1973). Matthews (1974) considered that a Lower Devonian thickness of about 4 km was not entirely improbable, and suggested that there might be a deep major fracture, not necessarily a thrust, along the southern margin of Exmoor, with parts of the Upper Palaeozoic succession thickening northwards from it.

4 A tectonic explanation in terms of a major southward-dipping thrust plane, with Carboniferous and Devonian rocks of relatively low density overthrust by the Devonian rocks of Exmoor, can account for the gravity anomaly and has also been invoked both to explain the near-juxtaposition of Namurian and Middle Devonian strata at Cannington Park (Whittaker, 1975) and to mark the 'Hercynian front', though a position on the north side of the Bristol Channel for the latter seems more probable (Matthews, 1974). Attempts to locate such a thrust plane by seismic methods have been unsuccessful (Brooks and James, 1975; E. M. Andrew, unpublished BGS survey).

5 Brooks and others (1977) have explained the Bouguer anomaly profile along longitude line 04°00'W in terms of a pre-Devonian basement of 6.1 to 6.3 km/s seismic velocity and 2.65 g/cm³ density which, overlain by Devonian and Carboniferous strata, rises northward across Exmoor to a minimum depth of about 2 km just off the north Devon coast and then falls again towards the Bristol Channel Syncline. An E–W seismic refraction line close to the north Devon coast established the basement velocity and provided depth control on the interpretation. Because of the featureless aeromagnetic field, the basal layer is also required to have low magnetic susceptibility. This interpretation fits the observed data well and does not depend on the choice of an arbitrary regional gradient; but it is probably geologically oversimplified (Whittaker, 1978) and an entirely convincing candidate for the basal seismic layer has yet to be found. (See also p. 58.)

Across and to the south of Exmoor, Bouguer anomaly values vary in a way reflecting changes in near-surface lithology. The flattening of the gradient over the Hangman Grits outcrop appears as a residual low after regional removal, offset to the south of the outcrop because of the southerly regional dip. A clearly defined low follows the outcrop of Pickwell Down Sandstones on both flanks of the Bray Valley Syncline [740 340]. Al-Sadi (1967) has interpreted the residual anomaly over the Pickwell Down Sandstones and has concluded that the southward dip is near 40° at the surface, decreasing to 10° or less at depth. An E–W trending low 4 km south of Barnstaple marks the Codden Hill Chert.

The aeromagnetic map is dominated by a linear anomaly belt which closely follows the geological strike and is confined to the upper parts of the Ilfracombe Slates and the overlying Morte Slates, both predominantly argillaceous formations. Inspection of the anomaly show that, though considerable variations of detail are apparent, there are several persistent components along most of its length. These are: a dominant positive anomaly, weakening towards the west, owing to a source extending to a considerable depth; two shorter-wavelength anomalies of lesser amplitude superimposed on the main anomaly, one near its crest and one about 800 m to the south; and a separate anomaly about 2.5 km south of the first, generally of low, variable amplitude and diminishing westwards except for a localised zone of enhanced intensity [51 43], near West Down. It appears that at least two structures are responsible for the anomaly belt. Ground magnetic surveys by Mr M. K. Lee revealed shorter-wavelength components, not resolved in the aeromagnetic survey, and showed that at least part of the anomaly was due to separate, elongated, near-surface bodies. The West Down anomaly was investigated on the ground and showed a predominantly positive anomaly of amplitude up to 250 nT due to a body up to 100 m wide at a few tens of metres depth.

Two boreholes at Honeymead Farm [7989 3935], to the east of the present district, were drilled to investigate the main anomaly; they showed that the magnetic mineral pyrrhotite was widespread as a coating on cleavage surfaces, though only locally abundant. Although remanence measurements were not made it is known that pyrrhotite can exhibit strong remanent and induced magnetisation (Cornwell, 1966), and it is therefore probable that variable disseminations of this mineral account for the anomaly belt. The local variations evident from the ground magnetic surveys may be partly due to variable weathering.

Computer modelling techniques have been used to get a better understanding of the magnetisation. It is found that the observed anomaly curves can be reproduced by 'thick sheet' models which extend from near surface to considerable, rather indeterminate, depth (generally greater than 2 km). Reasonable fits can be obtained for southward dips of 30° to 60°; corresponding values for the dip of the component of total (induced plus remanent) magnetisation perpendicular to strike are 80°S to 70°N for the southern structure, and 40°S to 70°S for the northern one. Figure 17 shows the fit obtained from one such model. If the dips of the two structures are the same, which is geologically reasonable, it appears impossible to construct a simple model without assuming different magnetisation directions for the two structures. The style of mineralisation suggests that the pyrrhotite was emplaced during a time of metamorphism and regional stress, perhaps during the Variscan earth movements; the different magnetisation directions imply that at least two phases of mineralisation occurred, separated by a time interval sufficient to allow considerable polar wandering, or that temperatures greater than the Curie point (about 300°C) acting subsequent to mineralisation have reset the magnetisation of one structure more than the other.

The regional dip of 30° or greater deduced from the modelling experiments is greater than the value of 10° to 15° deduced from geological considerations (see p. 58). A partial explanation may be that mineralisation followed cleavage rather than bedding (cleavage dips being generally the steeper); even so one might expect mineralisation to be confined to argillaceous formations, even at depth.   JMCT

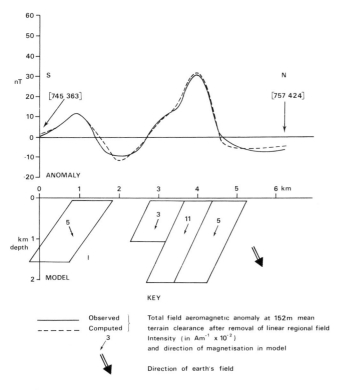

Figure 17   Interpretation of the Exmoor aeromagnetic anomaly

REFERENCES

AL-SADI, H. N. 1967. A gravity investigation of the Pickwell Down Sandstone, north Devon. *Geol. Mag.,* Vol. 104, pp. 63–72.

ARBER, E. A. N. 1904. The fossil flora of the Culm Measures of north-west Devon. *Proc. R. Soc. London,* Ser. B, Vol. 74, pp. 95–99.

— 1907. On the Upper Carboniferous rocks of west Devon and north Cornwall. *Q. J. Geol. Soc. London,* Vol. 63, pp. 1–28.

— 1911. *The coast scenery of north Devon.* (London: Dent.)

— and GOODE, R. H. 1915. On some fossil plants from the Devonian rocks of north Devon. *Proc. Cambridge Philos. Soc.,* Vol. 18, pp. 89–104.

ARBER, MURIEL A. 1960. Pleistocene sea levels in north Devon. *Proc. Geol. Assoc.,* Vol. 71, pp. 169–176.

— 1977. A brickfield yielding elephant remains at Barnstaple, north Devon. *Quaternary Newsletter,* No. 21, pp. 19–21.

AUSTIN, R. L., DRUCE, E. C., RHODES, F. H. T. and WILLIAMS, J. A. 1970. The value of conodonts in the recognition of the Devonian–Carboniferous boundary, with particular reference to Great Britain. *C. R. 6e Cong. Int. Stratigr. Géol. Carbonif. (Sheffield, 1967),* Vol. 2, pp. 431–443.

BADHAM, J. P. N., COSGROVE, M. E., CRUMP, D. R., EDWARDS, P. J., LEACH, A., LECLERE, M., SAUNDERS, R. A. and WINTER, A. 1979. Geochemical and ecological investigations at the South Molton Consols mine site. *Proc. Ussher Soc.,* Vol. 4, pp. 449–466.

BALCHIN, W. G. V. 1952. The erosion surfaces of Exmoor and adjacent areas. *Geogr. J.,* Vol. 118, pp. 453–476.

— 1966. The denudation chronology of south-west England. *In* Present views of some aspects of the geology of Cornwall and Devon. *R. Geol. Soc. Cornwall,* Commem. Vol. (for 1964), pp. 267–281.

BLYTH, F. G. H. 1957. The Lustleigh Fault in north-east Dartmoor. *Geol. Mag.,* Vol. 94, pp. 291–296.

BONNEY, T. G. 1878. Note on the felsite at Bittadon, north Devon. *Geol. Mag.,* Vol. 15, pp. 207–209.

BOTT, M. H. P., DAY, A. A. and MASSON SMITH, D. 1958. The geological interpretation of gravity and magnetic surveys in Devon and Cornwall. *Philos. Trans. R. Soc. London,* Ser. A, Vol. 251, pp. 161–191.

— and SCOTT, P. 1966. Recent geophysical studies in south-west England. *In* Present views of some aspects of the geology of Cornwall and Devon. *R. Geol. Soc. Cornwall,* Commem. Vol. (for 1964), pp. 25–44.

BOWEN, D. Q. 1969. A new interpretation of the Pleistocene succession in the Bristol Channel area. *Proc. Ussher Soc.,* Vol. 2, p. 86.

BRICE, D. 1970. Étude paléontologique et stratigraphique du Dévonien de l'Afghanistan. *Notes mém. Moyen-Orient,* Vol. 11, 364 pp.

BRISTOW, C. R. and COX, F. C. 1973. The Gipping Till: a reappraisal of East Anglian glacial stratigraphy. *J. Geol. Soc. London,* Vol. 129, pp. 1–37.

BROOKS, M., BAYERLY, M. and LLEWELLYN, D. J. 1977. A new geological model to explain the gravity gradient across Exmoor, north Devon. *J. Geol. Soc. London,* Vol. 133, pp. 385–393.

— and JAMES, D. G. 1975. The geological results of seismic refraction surveys in the Bristol Channel, 1970–1973. *J. Geol. Soc. London,* Vol. 131, pp. 163–182.

— and THOMPSON, M. S. 1973. The geological interpretation of a gravity survey of the Bristol Channel. *J. Geol. Soc. London,* Vol. 129, pp. 245–274.

BUTLER, D. E. 1981. Marine faunas from concealed Devonian rocks of southern England and their reflection of the Frasnian transgression. *Geol. Mag.,* Vol. 118, pp. 679–697.

COOK, A. H. and THIRLAWAY, H. I. S. 1952. A gravimeter survey in the Bristol and Somerset coalfields. *Q. J. Geol. Soc. London,* Vol. 107, pp. 255–286, 302–306.

CORNWELL, J. D. 1966. The magnetisation of Lower Carboniferous rocks from the north-west border of the Dartmoor granite, Devonshire. *Geophys. J.R. Astr. Soc.,* Vol. 12, pp. 381–403.

— 1971. Geophysics of the Bristol Channel area. *Proc. Geol. Soc. London,* No. 1664, pp. 286–289.

CURTIS, L. F. 1971. Soils of Exmoor Forest. *Soil Survey of England and Wales,* Spec. Surv. No. 5.

DAVIDSON, T. 1864–65. A monograph of the British Devonian Brachiopoda. *Palaeontogr. Soc.* (Monogr.).

DEARMAN, W. R. 1964. Wrench-faulting in Cornwall and south Devon. *Proc. Geol. Assoc.,* Vol. 74, pp. 265–287.

— 1971. A general view of the structure of Cornubia. *Proc. Ussher Soc.,* Vol. 2, pp. 220–236.

DE LA BECHE, H. T. 1839. Report on the geology of Cornwall, Devon and west Somerset. *Mem. Geol. Surv. G.B.*

DE RAAF, J. F. M., READING, H. G. and WALKER, R. G. 1965. Cyclic sedimentation in the lower Westphalian of north Devon, England. *Sedimentology,* Vol. 4, pp. 1–52.

DEWEY, H. 1910. Notes on some igneous rocks from north Devon. *Proc. Geol. Assoc.,* Vol. 21, p. 429.

— 1913. The raised beach of north Devon: its relation to others and to Palaeolithic Man. *Geol. Mag.,* Vol. 50, pp. 154–163.

— 1921. Lead, silver-lead and zinc ores of Cornwall, Devon and Somerset. *Spec. Rep. Miner. Resour. Mem. Geol. Surv. G.B.,* Vol. 21.

DINES, H. G. 1956. The metalliferous mining region of south-west England. *Mem. Geol. Surv. G.B.,* 2 Vols. 795 pp.

DIXON, E. E. L. and VAUGHAN, A. 1911. The Carboniferous succession in Gower (Glamorganshire); with notes on its fauna and conditions of deposition. *Q. J. Geol. Soc. London,* Vol. 67, pp. 477–571.

DONOVAN, D. T. 1971. In discussion of a symposium on the geology of the Bristol Channel. *Proc. Geol. Soc. London,* No. 1664, pp. 305–306.

EDMONDS, E. A. 1972a. The Pleistocene history of the Barnstaple area. *Rep. Inst. Geol. Sci.,* No. 72/2, 12 pp.

— 1972b. Drainage chronology of the South Molton area, Devonshire. *Bull. Geol. Surv. G.B.,* No. 42, pp. 99–104.

— 1974. Classification of the Carboniferous rocks of south-west England. *Rep. Inst. Geol. Sci.,* No. 74/13, 7 pp.

— McKEOWN, M. C. and WILLIAMS, M. 1969. South-west England. 3rd Edition. *Br. Reg. Geol., Inst. Geol. Sci.,* 130 pp.

— — — 1975. South-west England. 4th Edition. *Br. Reg. geol., Inst. Geol. Sci.,* 136 pp.

— WILLIAMS, B. J. and TAYLOR, R. T. 1979. Geology of Bideford and Lundy Island. *Mem. Geol. Surv. G.B.,* 143 pp.

— WRIGHT, J. E., BEER, K. E., HAWKES, J. R., WILLIAMS, M., FRESHNEY, E. C. and FENNING, P. J. 1968. Geology of the country around Okehampton. *Mem. Geol. Surv. G.B.,* 256 pp.

ETHERIDGE, R. 1867. On the physical structure of west Somerset and north Devon and on the palaeontological value of the Devonian fossils. *Q. J. Geol. Soc. London*, Vol. 23, pp. 568–698.

EVANS, J. W. 1922. The geological structure of the country around Combe Martin. *Proc. Geol. Assoc.*, Vol. 33, pp. 201–228.

— and POCOCK, R. W. 1912. The age of the Morte Slates. *Geol. Mag.*, Vol. 49, pp. 113–115.

— and STUBBLEFIELD, C. J. 1929. *Handbook of the geology of Great Britain*. (London: Thomas Murby & Co.)

FEDOROWSKI, J. 1965. Lindstroemiidae and Amplexocariniidae (Tetracoralla) from the Middle Devonian of Skaly, Holy Cross Mountains, Poland. *Acta Palaeont. Polonica*, Vol. 10, pp. 335–355, pls. 1–6.

FRESHNEY, E. C., BEER, K. E. and WRIGHT, J. E. 1979. Geology of the country around Chulmleigh. *Mem. Geol. Surv. G.B.*, 69 pp.

— EDMONDS, E. A., TAYLOR, R. T. and WILLIAMS, B. J. 1979. Geology of the country around Bude and Bradworthy. *Mem. Geol. Surv. G.B.*, 62 pp.

— and FENNING, P. J. 1967. The Petrockstow basin, N. Devon. *Proc. Ussher Soc.*, Vol. 1, pp. 278–281.

— and TAYLOR, R. T. 1971. The structure of mid-Devon and north Cornwall. *Proc. Ussher Soc.*, Vol. 2, pp. 241–248.

— — 1972. The Upper Carboniferous stratigraphy of north Cornwall and west Devon. *Proc. Ussher Soc.*, Vol. 2, pp. 464–471.

FRYER, G. 1960. Evolution of the land forms of Kerrier. *Trans. R. Geol. Soc. Cornwall*, Vol. 19, pp. 122–153.

GEORGE, T. N., HARLAND, W. B., AGER, D. V., BALL, H. W., BLOW, W. H., CASEY, R., HOLLAND, C. H., HUGHES, N. F., KELLAWAY, G. A., KENT, P. E., RAMSBOTTOM, W. H. C., STUBBLEFIELD, C. J. and WOODLAND, A. W. 1969. Recommendations on stratigraphical usage. *Proc. Geol. Soc. London*, No. 1656, pp. 139–166.

GOLDRING, R. 1955. The Upper Devonian and Lower Carboniferous trilobites of the Pilton Beds in N. Devon with an appendix on goniatites of the Pilton Beds. *Senckenb. Lethaea*, Vol. 36, pp. 27–48.

— 1957. The last toothed Productellinae in Europe (Brachiopoda, Upper Devonian). *Paläontol. Z.*, Vol. 31, pp. 207–228.

— 1962. The trace fossils of the Baggy Beds (Upper Devonian) of north Devon, England. *Paläontol. Z.*, Vol. 36, pp. 232–251.

— 1970. The stratigraphy about the Devonian–Carboniferous boundary in the Barnstaple area of north Devon, England. *C. R. 6e Congr. Int. Stratigr. Géol. Carbonif. (Sheffield, 1967)*, Vol. 2, pp. 807–816.

— 1971. Shallow-water sedimentation as illustrated in the Upper Devonian Baggy Beds. *Mem. Geol. Soc. London*, No. 5, 80 pp.

GREEN, G. W. 1955. North Exmoor floods, August 1952. *Bull. Geol. Surv. G.B.*, No. 7, pp. 68–84.

GREGORY, J. W. 1897. On the age of the Morte Slate fossils. *Geol. Mag.*, Vol. 34, pp. 59–62.

GROVES, A. W. 1940. Report on iron ores to the Home Ores Department, Ministry of Supply. (Unpublished.)

HALL, T. M. 1867. On the relative distribution of fossils throughout the North Devon Series. *Q. J. Geol. Soc. London*, Vol. 23, pp. 371–380.

— 1876. Fossil fish in north Devon. *Geol. Mag.*, Vol. 13, pp. 410–412.

HAMLING, J. G. 1908. Recently discovered fossils from the Lower and Upper Devonian beds of north Devon. *Trans. Devon. Assoc. Adv. Sci. Lit. Art*, Vol. 40, pp. 276–280.

— and ROGERS, I. 1910. Excursion to north Devon. *Proc. Geol. Assoc.*, Vol. 21, pp. 457–472.

HICKS, H. 1891. On the rocks of north Devon. *Q. J. Geol. Soc. London*, Vol. 47, pp. 7–11.

— 1896. On the Morte Slates and associated beds in north Devon and west Somerset. *Q. J. Geol. Soc. London*, Vol. 52, pp. 254–272.

HILL, D. 1939. Western Australian corals in the Wade Collection. *J. R. Soc. W. Aust.*, Vol. 25, pp. 141–152.

HIND, W. 1904. On the homotaxial equivalents of the Lower Culm of N. Devonshire. *Geol. Mag.*, Vol. 41, pp. 394–403, 584–587.

HINDE, G. J. and FOX, H. 1895. On a well-marked horizon of radiolarian rocks in the Lower Culm Measures of Devon, Cornwall and west Somerset. *Q. J. Geol. Soc. London*, Vol. 51, pp. 609–668.

HOLWILL, F. J. W. 1961. Limestones of the Ilfracombe Beds. Pp. 12–13 in *Abstr. Proc. Conf. Geol. Geomorphol. South-west Engl.* (Penzance: R. Geol. Soc. Cornwall.)

— 1963. The succession of limestones within the Ilfracombe Beds (Devonian) of north Devon. *Proc. Geol. Assoc.*, Vol. 73, pp. 281–293.

— 1964a. The Coral genus *Metriophyllum* Edwards and Haime. *Palaeontology*, Vol. 7, pp. 108–123.

— 1964b. The coral fauna from the Ilfracombe Beds of north Devon. *Proc. Ussher Soc.*, Vol. 1, pp. 126–129.

— 1968. Tabulate corals from the Ilfracombe Beds (Middle–Upper Devonian) of north Devon. *Palaeontology*, Vol. 11, Pt. 1, pp. 44–63.

— HOUSE, M. R., LANE, R., GAUSS, G. A., HENDRIKS, E. M. L. and DEARMAN, W. R. 1969. Summer (1966) field meeting in Devon and Cornwall. *Proc. Geol. Assoc.*, Vol. 80, pp. 43–62.

HOUSE, M. R. and SELWOOD, E. B. 1966. Palaeozoic palaeontology in Devon and Cornwall. *In* Present views of some aspects of the geology of Cornwall and Devon. *R. Geol. Soc. Cornwall*, Commem. Vol. (for 1964), pp. 45–86.

HUDSON, R. G. S. and COTTON, G. 1945. The Lower Carboniferous in a boring at Alport, Derbyshire. *Proc. Yorkshire Geol. Soc.*, Vol. 25, pp. 254–311.

INESON, P. R., MITCHELL, J. G. and ROTTENBURY, F. J. 1977. Potassium-argon isotopic age determinations from some north Devon mineral deposits. *Proc. Ussher Soc.*, Vol. 4, pp. 12–23.

INSTITUTE OF GEOLOGICAL SCIENCES. 1977. Quaternary map of the United Kingdom. South. 1:625 000.

JONES, O. T. 1931. Some episodes in the geological history of the Bristol Channel region. *Pres. Address to Sect. C. British Assoc.*, (Bristol), pp. 57–82.

— 1951. The drainage system of Wales and the adjacent regions. *Q. J. Geol. Soc. London*, Vol. 107, pp. 201–225.

KELLAWAY, G. A. 1971. Glaciation and the stones of Stonehenge. *Nature, London*, Vol. 232, pp. 30–35.

KIDSON, C. and WOOD, T. R. 1974. The Pleistocene stratigraphy of Barnstaple Bay. *Proc. Geol. Assoc.*, Vol. 85, pp. 223–237.

LANE, R. 1965. The Hangman Grits — an introduction and stratigraphy. *Proc. Ussher Soc.*, Vol. 1, pp. 166–167.

LECOMPTE, M. 1939. Les tabulés du Dévonien Moyen et Supérieur du bord sud du Bassin de Dinant. *Mem. Mus. R. d'Hist. Nat. Belg.*

LLOYD, A. J., SAVAGE, R. J. G., STRIDE, A. H. and DONOVAN, D. T. 1973. The geology of the Bristol Channel floor. *Philos. Trans. R. Soc. London*, Ser. A, Vol. 274, pp. 595–626.

LYSONS, D. and LYSONS, S. 1822. *Magna Britannia, Devonshire* 6. (London.)

MACKINTOSH, D. M. 1965. The tectonics of Namurian and Westphalian turbidite sandstones between Wanson Mouth and Rusey, north Cornwall. Unpublished PhD Thesis, University of Exeter.

MAILLIEUX, E. 1937. Les lamellibranches du Dévonien Inférieur de l'Ardenne. *Mem. Mus. R. d'Hist. Nat. Belg.*, No. 81.

MATTHEWS, S. C. 1974. Exmoor Thrust? Variscan Front? *Proc. Ussher Soc.*, Vol. 3, pp. 82–94.

MAW, G. 1864. On a supposed boulder clay in north Devon. *Q. J. Geol. Soc. London*, Vol. 20, pp. 445–451.

MILNE EDWARDS, H. and HAIME, J. 1850–55. A monograph of the British fossil corals. *Palaeontogr. Soc.*, (Monogr.).

MITCHELL, G. F. 1960. The Pleistocene history of the Irish Sea. *Adv. Sci. London*, Vol. 17, pp. 313–325.

— 1972. The Pleistocene history of the Irish Sea; second approximation. *Sci. Proc. R. Dublin Soc.*, Ser. A, Vol. 4, pp. 181–190.

MOORBATH, S. 1962. Lead isotope abundance studies on mineral occurrences in the British Isles and their geological significance. *Philos. Trans. R. Soc. London*, Ser. A, Vol. 254, pp. 295–360.

MOORE, E. W. J. 1929. The occurrence of *Reticuloceras reticulatum* in the Culm of north Devon. *Geol. Mag.*, Vol. 66, pp. 356–358.

MUIR WOOD, H. and COOPER, G. A. 1960. Morphology, classification and life habits of the Productoidea (Brachiopoda). *Mem. Geol. Soc. Am.*, Vol. 81, 447 pp.

ORCHARD, M. J. 1979. On a *varcus* Zone conodont fauna from the Ilfracombe Slates (Devonian, north Devon). *Geol. Mag.*, Vol. 116, pp. 129–134.

ORME, A. R. 1962. Abandoned and composite sea cliffs in Britain and Ireland. *Ir. Geogr.*, Vol. 4, pp. 279–291.

ORWIN, C. S. 1929. *The reclamation of Exmoor Forest.* (London.)

PARTRIDGE, E. M. 1902. *Echinocaris whidbornei* (Jones and Woodward) and *Echinocaris sloliensis*, n. sp. *Geol. Mag.*, Vol. 39, pp. 307–308.

PATTISON, S. R. 1865. A day in the north Devon mining district. *Trans. R. Geol. Soc. Cornwall*, Vol. 7, pp. 223–227.

PAUL, H. 1937. The relationship of the Pilton Beds in north Devon to their equivalents on the continent. *Geol. Mag.*, Vol. 74, pp. 433–442.

PHILLIPS, J. 1841. *Figures and descriptions of the Palaeozoic fossils of Cornwall, Devon and west Somerset.* (London: Longman, Brown, Green and Longmans.)

PRANTL, F. 1938. Some Laccophyllidae from the Middle Devonian of Bohemia. *Ann. Mag. Nat. Hist.*, Vol. 11, Pt. 2, pp. 18–41.

PRENTICE, J. E. 1960a. Dinantian, Namurian and Westphalian rocks of the district south-west of Barnstaple, north Devon. *Q. J. Geol. Soc. London*, Vol. 115, pp. 261–289.

— 1960b. The stratigraphy of the Upper Carboniferous rocks of the Bideford region, north Devon. *Q. J. Geol. Soc. London*, Vol. 116, pp. 397–408.

READING, H. G. 1965. Recent finds in the Upper Carboniferous of south-west England and their significance. *Nature, London*, Vol. 208, pp. 745–748.

REED, F. R. C. 1943. Notes on certain Upper Devonian brachiopods figured by Whidborne. *Geol. Mag.*, Vol. 80, pp. 69–78, 95–106, 132–138.

— 1944. Notes on the Upper Devonian trilobites in the Whidborne Collection in the Sedgwick Museum. *Geol. Mag.*, Vol. 81, pp. 121–126.

RICHTER, R. and RICHTER, E. 1939. Proetidae von oberdevonischer Tracht im deutschen, englishchen und mittelmeerischen Unter-Karbon. *Senckenbergiana*, Vol. 21, pp. 82–112.

RISDON, T. 1811. *A survey of the county of Devon.* (London.)

ROGERS, I. 1907. On fossil fish. *Trans. Devon. Assoc. Adv. Sci. Lit. Art*, Vol. 39, pp. 394–398.

— 1908. On the submerged forest at Westward Ho! A history of Northam Burrows. *Trans. Devon. Assoc. Adv. Sci. Lit. Art*, Vol. 40, pp. 249–259.

— 1909. On a further discovery of fossil fish and mollusca in the Upper Culm Measures of north Devon. *Trans. Devon. Assoc. Adv. Sci. Lit. Art*, Vol. 41, pp. 309–319.

— 1910. A synopsis of the fossil flora and fauna of the Upper Culm Measures of north-west Devon. *Trans. Devon. Assoc. Adv. Sci. Lit. Art*, Vol. 42, pp. 538–564.

— 1919. Fossil fishes in the Devonian of north Devon. *Geol. Mag.*, Vol. 56, pp. 100–101.

— 1926. On the discovery of fossil fishes and plants in the Devonian rocks of north Devon. *Trans. Devon. Assoc. Adv. Sci. Lit. Art*, Vol. 58, pp. 223–234.

ROTTENBURY, F. J. 1974. Geology and mining history of the metalliferous mining areas of Exmoor. Unpublished Ph.D Thesis, University of Leeds.

SCRIVENER, R. C. and BENNETT, M. J. 1980. Ore genesis and controls of mineralisation in the Upper Palaeozoic rocks of north Devon. *Proc. Ussher Soc.*, Vol. 5, pp. 54–58.

SCRUTTON, C. T. 1968. Colonial Phillipsastraeidae from the Devonian of south-east Devon, England. *Bull. Brit. Mus. Nat. Hist. (Geol.)*, Vol. 15, pp. 183–281.

SEDGWICK, A. and MURCHISON, R. I. 1836. A classification of the Old Slate Rocks of the north of Devonshire. *Rep. Br. Assoc. Adv. Sci.*, No. 5, p. 95.

— — 1837. On the physical structure of Devonshire, and on the subdivisions and geological relations of its older stratified deposits. *Proc. Geol. Soc. London*, No. 2, p. 559.

— — 1840. On the physical structure of Devonshire, and on the subdivisions and geological relations of its older stratified deposits. *Trans. Geol. Soc. London*, 2nd Ser., Vol. 5, pp. 633–705.

SHEARMAN, D. J. 1962. Aspects of the geology of the Ilfracombe Beds (Devonian) of north Devon; structure and lithological succession. *Geol. Assoc.*, Circ. No. 641.

— 1967. On Tertiary fault movements in north Devonshire. *Proc. Geol. Assoc.*, Vol. 78, pp. 555–566.

SIMPSON, S. 1951. Some solved and unsolved problems of the stratigraphy of the marine Devonian in Great Britain. *Abs. Senckenb. naturforsch. Ges.*, Vol. 485, pp. 53–66.

— 1957. On the trace-fossil *Chondrites*. *Q. J. Geol. Soc. London*, Vol. 112, pp. 475–496.

— 1964. The Lynton Beds of north Devon. *Proc. Ussher Soc.*, Vol. 1, pp. 121–122.

— 1971. The Variscan structure of north Devon. *Proc. Ussher Soc.*, Vol. 2, pp. 249–252.

SMYTH, W. W. 1859. On the iron ores of Exmoor. *Q. J. Geol. Soc. London*, Vol. 15, pp. 105–109.

STEPHENS, N. 1961. A re-examination of some Pleistocene sections in Cornwall and Devon. Pp. 21–23 in *Abstr. Proc. Conf. Geol. Geomorphol. South west Engl.* (Penzance: R. Geol. Soc. Cornwall.)

— 1966. Some Pleistocene deposits in north Devon. *Biul. Periglacjalny*, No. 8, pp. 103–114.

— 1970. The west country and southern Ireland. In *The glaciations of Wales and adjoining regions*. LEWIS, C. A. (Editor). (London: Longmans).

TAYLOR, C. W. 1956. Erratics of the Saunton and Fremington areas. *Trans. Devon. Assoc. Adv. Sci. Lit. Art*, Vol. 88, pp. 52–64.

THOMAS, A. N. 1940. The Triassic rocks of north-west Somerset. *Proc. Geol. Assoc.*, Vol. 51, pp. 1–43.

TSIEN, H. H. 1970. Espèces du genre *Disphyllum* (Rugosa) dans le Dévonien Moyen et le Frasnien de la Belgique. *Ann. Soc. Géol. Belg.*, Vol. 93, pp. 159–182.

TUNBRIDGE, I. P. 1976. Notes on the Hangman Sandstones (Middle Devonian) of north Devon. *Proc. Ussher Soc.*, Vol. 3, p. 339.

— 1978. *In* A field guide to selected areas of the Devonian of south-west England. International symposium on the Devonian System (P.A.D.S. 78). *Palaeont. Assoc.*, pp. 11–13.

USSHER, W. A. E. 1879. On the geology of parts of Devon and west Somerset, north of South Molton and Dulverton. *Proc. Somerset Arch. Nat. Hist. Soc.*, Vol. 5, pp. 1–20.

— 1881. On the Palaeozoic rocks of north Devon and west Somerset. *Geol. Mag.*, Vol. 18, pp. 441–448.

— 1887. The Culm of Devonshire. *Geol. Mag.*, Vol. 24, pp. 10–17.

— 1889. The Triassic rocks of west Somerset and the Devonian rocks on their borders. *Proc. Somerset Arch. Nat. Hist. Soc.*, Vol. 15, pp. 1–36.

— 1892. The British Culm Measures. *Proc. Somerset Arch. Nat. Hist. Soc.*, Vol. 38, pp. 111–219.

— 1900. The Devonian, Carboniferous and New Red rocks of west Somerset, Devon and Cornwall. *Proc. Somerset Arch. Nat. Hist. Soc.*, Vol. 46, pp. 1–64.

— 1901. The Culm-Measure types of Great Britain. *Trans. Inst. Min. Eng.*, Vol. 20, pp. 360–391.

— 1906. *Devonshire: Victoria History of the Counties of England.* (London.)

— 1913. Geology of the country around Newton Abbot. *Mem. Geol. Surv. G.B.*, 149 pp.

— and CHAMPERNOWNE, A. 1879. Notes on the structure of the Palaeozoic districts of west Somerset. *Q. J. Geol. Soc. London*, Vol. 35, pp. 532–548.

VALPY, R. H. (anon.) 1867. *Notes on the geology of Ilfracombe.* (Twiss and Sons.)

VANDERCAMMEN, A. 1959. Essai d'etude statistique des *Cyrtospirifer* du Frasnian de la Belgique. *Mem. Inst. R. Sci. Nat. Belg.*, Vol. 145.

VAUGHAN, A. 1904. Note on the Lower Culm of north Devon. *Geol. Mag.*, Vol. 41, pp. 530–532.

WEBBY, B. D. 1963. Written contribution to discussion of a paper read 6 April 1962 (Holwill, 1963). *Proc. Geol. Assoc.*, Vol. 74, pp. 261–262.

— 1964. Devonian corals and brachiopods from the Brendon Hills, west Somerset. *Palaeontology*, Vol. 7, pp. 1–22.

— 1965. The stratigraphy and structure of the Devonian rocks in the Brendon Hills, west Somerset. *Proc. Geol. Assoc.*, Vol. 76, pp. 39–60.

— 1966a. The stratigraphy and structure of the Devonian rocks in the Quantock Hills, west Somerset. *Proc. Geol. Assoc.*, Vol. 76, pp. 321–343.

— 1966b. Middle–Upper Devonian palaeogeography of north Devon and west Somerset, England. *Palaeogeog., Palaeoclim., Palaeoecol.*, Vol. 2, pp. 27–46.

WHIDBORNE, G. F. 1896. A preliminary synopsis of the fauna of the Pickwell Down, Baggy and Pilton Beds. *Proc. Geol. Assoc.*, Vol. 14, pp. 371–377.

— 1896–1907. A monograph of the Devonian fauna of the south of England, 3. The fauna of the Marwood and Pilton Beds of north Devon and Somerset. *Palaeontogr. Soc.* (Monogr.)

— 1901. Devonian fossils from Devonshire. *Geol. Mag.*, Vol. 38, pp. 529–539.

WHITTAKER, A. 1975. Namurian strata near Cannington Park, Somerset. *Geol. Mag.*, Vol. 112, pp. 325–326 (Correspondence).

— 1976. Notes on the Lias outlier near Selworthy, west Somerset. *Proc. Ussher Soc.*, Vol. 3, pp. 355–359.

— 1978. Discussion of the gravity gradient across Exmoor, north Devon. *J. Geol. Soc. London*, Vol. 135, pp. 353–354.

WILLIAMS, D. 1837. On some fossil wood etc. *Rep. Br. Assoc. Adv. Sci.*, No. 6, p. 94.

APPENDIX

List of Geological Survey photographs

Copies of these photographs may be seen in the library of the British Geological Survey, Exhibition Road, South Kensington, London SW7 2DE. Prints and lantern slides may be bought. Photographs with numbers up to 8740 were taken by Mr J. Rhodes and are available in black and white only. Those with higher numbers were taken by Messrs C. Jeffery and H. J. Evans, and are obtainable in colour and black and white. The photographs belong to Series A.

ILFRACOMBE (277) SHEET

General views

5975	Coast scenery due to erosion of the Ilfracombe Slates
5980–1	Coast scenery of Ilfracombe Slates
5988	Headland of Ilfracombe Slates
5992	Combe Martin Bay and Little Hangman, Ilfracombe
5993, 5995	Little Hangman, Combe Martin
5997	Great Hangman. Typical 'hog's back' cliff
5998	V-shaped valley cut along the strike of the Hangman Grits
5999–6001	Coast scenery. Woody Bay, Crock Point and Foreland Point
6007	Valley of Rocks, Lynton
6008	Castle Rock and Duty Point, Valley of Rocks, Lynton
6009–13	Castle Rock and Valley of Rocks area, Lynton
6019	Foreland Point and Countisbury Beacon
6020	Hollerday Hill, 'hog's back' cliff, Lynton
13032	Topography of Exmoor [691 440]
13033	Topography of the Ilfracombe area, Torrs Park, Ilfracombe [515 465] (Plate 5)

Devonian

5972, 5984	Quartz veins in the Ilfracombe Slates
5973	Cliffs of calcareous slates of the Ilfracombe Slates
5974	Highly folded Devonian beds
5976	Cleaved and folded Ilfracombe Slates
5977	Pinnacles due to weathering of Ilfracombe Slates limestones
5978–9	Limestone weathering into sea stacks
5982	Folds passing into shear-lenticles, Ilfracombe Slates
5983	Folded Ilfracombe Slates
5985	Cliffs and reefs of Ilfracombe Slates
5996	Contorted slates, Ilfracombe Slates
6003	Reefs of Lynton Slates
6017	Cleaved and sheared Devonian beds
6018	Cliff of Lynton Slates
6021–2	Asymmetric folds near junction of Foreland Grits and Lynton Slates
6023	Devonian grits
13018	Bedding plane slip in Hangman Grits, Heddon's Mouth [655 497]
13019	Quartz veins in Hangman Grits, Heddon's Mouth [655 497]
13020	Sedimentary structures in Hangman Grits, Heddon's Mouth [655 497]. (Plate 2)

13021–3	Coastal exposures in Ilfracombe Slates, Combe Martin [576 475; 576 474; 570 474]. (13023 — Plate 3)
13024	David's Stone Limestone, Sandy Bay, Combe Martin (type locality) [571 474]
13025	Inverted David's Stone Limestone, Sandy Bay, Combe Martin [570 473]
13026–7	Jenny Start Limestone, Jenny Start Point [566 478]
13028	Combe Martin Slates, Hele Bay, Ilfracombe [541 485]
13029	Cliffs of Combe Martin Slates, Hele Bay, Ilfracombe [537 479]
13030	Tectonic structures in Combe Martin Slates, Hele Bay, Ilfracombe [537 479]
13031	Structure and lithologies of Kentisbury Slates, Ilfracombe [519 479]. (Plate 1)

Pleistocene and Recent

5986	Coast erosion of Ilfracombe Slates
5987	Reef and caves in Ilfracombe Slates
5989–90	Drowned river mouth, Ilfracombe
5991	Cave eroded along a fault, Ilfracombe
5994	Lester Point. Marine erosion of Ilfracombe Slates
6002	Marine erosion of sheared and inclined Lynton Slates
6004	Weathering of Lynton Slates
6005	Natural arch in Lynton Slates
6006	Coastal valleys breached by sea and headlands
6014–6	Pinnacles and crags, weathering of Lynton Slates
6024–6	Dry valley, Ilfracombe

[Photographs nos. 8667–8740 relate to the Exmoor floods of 1952]

8667	Ravines and potholes in moorland track, exposing Head
8668	Uppermost reaches of the West Lyn River
8678	Cliffs in alluvium, showing boulder-filled channel
8679	Abandoned river channel now filled with boulders
8680–1	Breaches in stone wall
8682	New river channel being cut back in alluvial gravel
8683	Ravines and potholes associated with a small stream
8684	Section in Head, 2 m high
8685	Widened stream bed showing boulder alluvium
8686	Flood-plain deposition
8687	Enlarged river bed
8688	Straightened river course
8689	Undercut cliff of Head on Hangman Grits
8690	Increased erosion within meander, with consequent destruction of barn
8691	Boulders and trees on alluvial surface, held up by ledge and wall
8692–3	Newly cut gorge 2 m deep
8694, 8696, 8701–6	West Lyn River gorge, including various boulder beds and landslips
8695	Head on Lynton Slates in West Lyn River gorge
8697	Chipping of Lynton Slates in river bed
8698–9	Bridge built on solid rock
8700	Erosion of Head, concrete and masonry, exposing Lynton Slates

8707	Cliff of Head on Lynton Slates
8708–9	Eroded flood debris
8710, 8717	Boulder-choked bed of East Lyn River gorge
8711	Cliff of Head
8712	Chipping of Devonian rocks
8713	East Lyn River gorge above Lynmouth
8714–6, 8718–9, 8725,	Houses and flood damage
8729–31	Gravel and boulder deposits, Farley
8726	Enlarged river course, Farley
8727	Gravel fan at base of hillside ravine in Head, Farley
8728	Straightened river course, Farley
8732	Gravel fan, Farley
8733	Gravel deposited behind an obstacle, Cheriton
8734	Flood-damaged farmhouse, Cheriton
8735	Boulder spread in river diversion, Cheriton
8736	Gouging and demolished stone wall, Roborough Castle, Cheriton
8737–8	Middle reaches of Hoaroak Water, Cheriton
8739	Choked abandoned river course, Cheriton
8740	River diversion, Cheriton

BARNSTAPLE (293) SHEET

General views

5964	View across Carboniferous country, Tower Hamlet Quarry, Bishop's Tawton
5967	'Hog's back' hill of radiolarian chert, from Hearson Hill, Swimbridge
11832–4	View north-west (11832), south (11833) and east (11834) from Codden Hill [582 296]
11835–6	Carboniferous topography, Filleigh [6630 2815; 6340 2950]. (11836 — Plate 9)
11841	View from Venn Quarries, Barnstaple [583 300]
11842	Crackington Formation topography, Grilstone Farm [730 247]
11850	Estuary of the River Taw, Bickington [537 332]. (Plate 11)
11851	Estuary of the River Taw, Chivenor [5096 3440]
11852–3	Drift topography, Filleigh [6735 2786; 6850 2775]
13008, 13012	River Barle, Simonsbath [763 387; 748 386]. (13012 — Plate 6)
13009	Challacombe Reservoir [696 422]
13010	Pinkworthy Pond, Exmoor [723 423]. (Plate 4)
13011	River Exe, Exmoor [764 416]

Devonian

13013	Morte Slates, Simonsbath [7613 3850]
13014	Pilton Shales, Brayford [748 386]
13015–7	Sandstones in Pilton Shales, Brayford [7613 3850; 6898 3336; 692 329]. (13016 — Plate 7)

Carboniferous

5960–1	Steeply inclined radiolarian chert, Codden Hill Quarry
5962	Sigmoidal folding in radiolarian chert, Codden Hill Quarry
5963	Vertical radiolarian chert, Tower Hamlet Quarry, Bishop's Tawton
5965	Ridge of radiolarian chert, Tower Hamlet Quarry, Bishop's Tawton
11823	Quarry in Lower Carboniferous beds, Hele, Barnstaple [5424 3226]

11824	Lower Carboniferous cherts, Templeton Quarry, Eastacombe, Barnstaple [544 297]
11825	Axis of anticline, Templeton Quarry, Eastcombe, Barnstaple [544 297]
11826–7	Lower Carboniferous cherts, Codden Hill Quarry [569 297]. (11827 — Plate 8)
11828	Old Limestone quarry, Venn, Barnstaple [5808 3091]. (Plate 12)
11829, 11831	Lower Carboniferous cherts, Swimbridge [6156 3028; 6200 3015]
11830	Folding in Lower Carboniferous cherts, Swimbridge [6210 2834]
11837	Paralic sandstone in Crackington Formation, Langley Barton [5637 2486]
11838–40	Venn Quarries, Barnstaple [581 302; 580 306; 5792 3055]. (11839 — Plate 13)

Pleistocene and Recent

[Photographs nos. 8660 to 8783 relate to the Exmoor floods of 1952]

8660	Junction of two source streams of the Thornworthy brook
8661	Western source stream of the Thornworthy brook
8662	Scouring of slates, Thornworthy brook
8663	Peat rafts carried by flood water, Thornbury brook
8668	Uppermost reaches of the West Lyn River
8669	Landslips in peat, West Lyn River
8670	Small landslips in Head in Gullies, West Lyn River
8671	Head 4.5 m on slate and sandstone, exposed by intensive downcutting, side stream, West Lyn River
8672–4	Ravines in peat and Head, side stream, West Lyn River
8675	Widened river bed, West Lyn River
8676–7	Boulders and gravel deposits, West Lyn River
8720	Upper reaches of Hoaroak Water, near Hoar Oak Tree
8721	Ravines and landslips in Head on Ilfracombe Slates, Hoaroak Water
8722	Gravel spreads on alluvium, Hoaroak Water, near Hoar Oak Tree
8723	Hillside ravine of temporary stream, near Hoar Oak Tree
8724	Flood debris covering alluvium, near Hoar Oak Tree
8741	View downstream, River Bray, Challacombe
8742	Ravines exposing Ilfracombe Slates, River Bray, Challacombe
8743	Ravines exposing Head on slates, River Bray, Challacombe
8744–6	Ravines, River Bray, Challacombe
8747	Low-angle creep of peat on slaty Head, River Bray, Challacombe
8748	Ilfracombe Slates exposed by slip, River Bray, Challacombe
8749	River Bray gravel deposits, Swincombe Rocks, Challacombe
8750	Challacombe Reservoir and slate-gravel delta
8751	Slate-gravel, River Bray, Challacombe
8752	Boulder spread, Yarbury Combe, Challacombe
8753	Erosion in Head and Ilfracombe Slates, Yarbury Combe, Challacombe
8754	Waterfall, Yarbury Combe, Challacombe
8755–6	Downcutting in Head on slates, Yarbury Combe, Challacombe
8757–8	Boulder spread, Yarbury Combe, Challacombe

8759–68	Temporary flood-water stream, Challacombe–Simonsbath road
8769–70	Hillside erosion, Challacombe–Simonsbath road
8771–2	Erosion in peat on Head, side stream of River Barle, Simonsbath
8773–4	Trenching in slaty Head, Simonsbath
8775–7	Erosion, Tangs Bottom, Simonsbath
8778	Landslips in Head on Ilfracombe Slates, Simonsbath
8779, 8781	Peat balls formed by stream action, Simonsbath
8780	Slaty gravel and peat balls, side stream of River Barle, Simonsbath
8782	River Barle, Simonsbath–Kinsford Gate road
8783	White Water, Clovenrocks Bridge, Simonsbath
11843	Higher Gorse Claypits, Fremington [5297 3167]
11844–5	Pebbly drift, Chivenor [5096 3440]
11846–7	Pebbly drift, Lake, Barnstaple [5534 3150]. (11846 — Plate 10)
11848–9	Pebbly drift, Penhill Point [517 340]

INDEX OF FOSSILS

GENERAL INDEX